SpringerBriefs in Mathematics

SpringerBriefs in Mathematics showcases expositions in all areas of mathematics and applied mathematics. Manuscripts presenting new results or a single new result in a classical field, new field, or an emerging topic, applications, or bridges between new results and already published works, are encouraged. The series is intended for mathematicians and applied mathematicians. All works are peer-reviewed to meet the highest standards of scientific literature.

Titles from this series are indexed by Scopus, Web of Science, Mathematical Reviews, and zbMATH.

More information about this series at http://www.springer.com/series/10030

Adam Kubica • Katarzyna Ryszewska •
Masahiro Yamamoto

Time-Fractional Differential Equations

A Theoretical Introduction

 Springer

Adam Kubica
Warsaw University of Technology
Warszawa, Poland

Katarzyna Ryszewska
Warsaw University of Technology
Warszawa, Poland

Masahiro Yamamoto
Graduate School of Mathematical Sciences
The University of Tokyo
Tokyo, Japan

ISSN 2191-8198   ISSN 2191-8201   (electronic)
SpringerBriefs in Mathematics
ISBN 978-981-15-9065-8   ISBN 978-981-15-9066-5   (eBook)
https://doi.org/10.1007/978-981-15-9066-5

Mathematics Subject Classification: 35R11, 26A33

This Springer imprint is published by the registered company Springer Nature Singapore Pte Ltd.
The registered company address is: 152 Beach Road, #21-01/04 Gateway East, Singapore 189721, Singapore

Preface

Recently, fractional differential equations have attracted great attention and many studies have been performed. However, there are not many works that cover the theory of partial differential equations, so that unnecessarily lengthy arguments and also arguments that lack rigor are sometimes presented, and it may be difficult to gain unified views of the related fields.

This concise book provides rigorous treatments for time-fractional derivatives in Sobolev spaces and solutions to initial boundary value problems for time-fractional partial differential equations and establishes the foundation of the theory for fractional differential equations. The results here should be fundamental also for discussing other important topics such as nonlinear dynamical systems, optimal control, and inverse problems for fractional differential equations. Although our approach can work for more general fractional differential equations, focusing on a differential equation with a single time-fractional derivative, we describe the theory.

The parts of the book have been presented as graduate courses at Sapienza University of Rome and The University of Tokyo. The authors thank Professor Bangti Jin (University College London) for valuable comments.

The first and second authors are supported partly by the National Science Centre, Poland through 2017/26/M/ST1/00700 Grant, and the third author is partly supported by Grant-in-Aid for Scientific Research (S) 15H05740 of Japan Society for the Promotion of Science and by the National Natural Science Foundation of China (nos. 11771270 and 91730303) and the "RUDN University Program 5–100."

Warszawa, Poland
Warszawa, Poland
Tokyo, Japan
November 2020

Adam Kubica
Katarzyna Ryszewska
Masahiro Yamamoto

Remarks for Readers

- In this book, we minimize the references. In Chap. 2, we use some basic knowledge of functional analysis and operator theory. The readers can consult suitable textbooks on functional analysis, and here we rely on Brezis [5], Kato [13], Pazy [23], and Tanabe [27]. In particular, for the function space, we refer to Adams [2]. Needless to say, there are other reputable source books.
- The readers who wish to use this book for quickly understanding a theoretical essence of fractional differential equations are encouraged to skip the proofs of Lemmata 2.2–2.4 and Theorem 2.1. They can skip also Sect. 2.6 of Chap. 2 and Sects. 5.2 and 5.3 of Chap. 5.
- We use the following numbering: for example, Theorem 2.1 means the first theorem in Chap. 2.

Contents

Chapter 1
Basics on Fractional Differentiation and Integration

Derivatives of orders of natural numbers have been widely known and commonly used since the origin of the calculus in the seventeenth century. On the other hand, derivatives whose orders are not necessarily natural numbers seem not to be well-known but have been considered since Leibniz (e.g., Ross [25]). Moreover, we can refer to a historical paper [1] by N.H. Abel. He discussed the following problem:

Given a function $f(t)$, find a function $v(t)$ such that

$$\frac{1}{\sqrt{\pi}} \int_0^t (t-s)^{-1/2} v(s) ds = f(t), \quad 0 \le t \le T. \tag{1.1}$$

This is an integral equation and appears to be related to the following mechanical problem: we consider that a mass goes along a curve on the vertical surface only by gravity starting from a point O to the end point P, and assume that the travel time is given by a function $f(t)$, where t is the distance in the vertical direction from O. Then, determine a shape of such a curve realizing the traveling time $f(t)$.

Abel established that for given $f \in C^1[0, T]$, the solution v to (1.1) exists uniquely and is given by

$$v(t) = \frac{1}{\sqrt{\pi}} \frac{d}{dt} \int_0^t (t-s)^{-1/2} f(s) ds, \quad 0 \le t \le T, \qquad \beta > 0. \tag{1.2}$$

Here and henceforth by $\Gamma(\cdot)$ we denote the gamma function:

$$\Gamma(\beta) = \int_0^\infty e^{-t} t^{\beta-1} dt, \quad \beta > 0.$$

We note that $\Gamma(n+1) = n! := n(n-1) \cdots 3 \cdot 2 \cdot 1$ for $n \in \mathbb{N}$.

We can regard (1.2) as the half-order derivative of f as follows. We note that $\frac{1}{\Gamma(n+1)} \int_0^t (t-s)^n v(s) ds$ is a function which is obtained by $(n+1)$-times

© The Author(s), under exclusive licence to Springer Nature Singapore Pte Ltd. 2020
A. Kubica et al., *Time-Fractional Differential Equations*, SpringerBriefs in Mathematics, https://doi.org/10.1007/978-981-15-9066-5_1

integrating $v(t)$. Thus we can interpret $\frac{1}{\sqrt{\pi}} \frac{d}{dt} \int_0^t (t-s)^{-1/2} f(s) ds$ as a function which is differentiated $\frac{1}{2} = -\frac{1}{2} + 1$-times. That is, setting $n = -\frac{1}{2}$ formally in $\frac{1}{\Gamma(n+1)} \int_0^t (t-s)^n v(s) ds$ and noting that $\sqrt{\pi} = \Gamma(1/2)$, we see that the right-hand side of (1.2) means the half-times derivative of f. Thus we can interpret that (1.2) means that the half-order derivative of f is v when f is the half-times integration of v, and establishes the inversion of the half-order integral in terms of the half-order derivative.

As for other applications of fractional derivatives, we refer to Gorenflo and Vessella [8].

Although it is later seen that the rigorous treatments are necessary, in this chapter we intuitively discuss fractional calculus as an introduction.

Throughout the book we assume

$$0 < \alpha < 1.$$

In this chapter, as functions we mainly consider $u \in L^1(0, T)$ satisfying $\frac{du}{dt} \in L^1(0, T)$, if not specified. We set

$$D_t^\alpha u(t) := \frac{1}{\Gamma(1-\alpha)} \frac{d}{dt} \int_0^t (t-s)^{-\alpha} u(s) ds \qquad (1.3)$$

and

$$d_t^\alpha u(t) := \frac{1}{\Gamma(1-\alpha)} \int_0^t (t-s)^{-\alpha} \frac{du}{ds}(s) ds. \qquad (1.4)$$

We call D_t^α and d_t^α the Riemann–Liouville derivative and the Caputo derivative respectively. The notations of the derivatives are different from e.g., Kilbas et al. [15], Podlubny [24], and in Chap. 2, the fractional derivative ∂_t^α in fractional Sobolev spaces is defined and plays an essential role rather than the classical Riemann–Liouville derivative and the Caputo derivative.

Throughout this book, we interpret equalities in variables t, x, etc., as that they hold for almost all t, x in domains under consideration, and we do not specify if there is no fear of confusion.

In this book, we mainly consider the case $0 < \alpha < 1$, but for general $\alpha > 0$, $\notin \mathbb{N}$, choosing $m \in \mathbb{N}$ satisfying $m - 1 < \alpha < m$, for $u \in C^m[0, T]$, we can define d_t^α and D_t^α by

$$d_t^\alpha u(t) = \frac{1}{\Gamma(m-\alpha)} \int_0^t (t-s)^{m-\alpha-1} \frac{d^m u}{ds^m}(s) ds$$

and

$$D_t^\alpha u(t) = \frac{1}{\Gamma(m-\alpha)} \frac{d^m}{dt^m} \int_0^t (t-s)^{m-\alpha-1} u(s) ds.$$

Moreover, for $\alpha > 0$, we define the Riemann–Liouville fractional integral operator by

$$J^\alpha f(t) = \frac{1}{\Gamma(\alpha)} \int_0^t (t-s)^{\alpha-1} f(s) ds, \tag{1.5}$$

which we can regard as an operation of α-times integration.

Example 1.1 Let $0 < \alpha < 1$ and $\beta > 0$. Then t^β, $\frac{dt^\beta}{dt} = \beta t^{\beta-1} \in L^1(0, T)$ and we have

$$d_t^\alpha t^\beta = D_t^\alpha t^\beta = \frac{\Gamma(\beta+1)}{\Gamma(1-\alpha+\beta)} t^{\beta-\alpha}. \tag{1.6}$$

In fact, this follows from the classical formula:

$$\int_a^b (t-a)^{p-1}(b-t)^{q-1} dt = \frac{\Gamma(p)\Gamma(q)}{\Gamma(p+q)}(b-a)^{p+q-1}, \quad p, q > 0, \ a < b. \tag{1.7}$$

We can prove also:

Lemma 1.1 *Let $u \in C^1[0, T]$. Then for $0 < \alpha < 1$, we have*

$$d_t^\alpha u(t) + \frac{u(0)}{\Gamma(1-\alpha)} t^{-\alpha} = D_t^\alpha u(t), \quad 0 \le t \le T. \tag{1.8}$$

Proof Since $u \in C^1[0, T]$, integration by parts yields

$$\int_0^t (t-s)^{-\alpha} u(s) ds = \left[\frac{u(s)(t-s)^{1-\alpha}}{1-\alpha} \right]_{s=t}^{s=0} + \frac{1}{1-\alpha} \int_0^t (t-s)^{1-\alpha} \frac{du}{ds}(s) ds$$

$$= \frac{u(0)}{1-\alpha} t^{1-\alpha} + \frac{1}{1-\alpha} \int_0^t (t-s)^{1-\alpha} \frac{du}{ds}(s) ds.$$

By differentiating both sides, the proof is complete by $1 - \alpha > 0$. ∎

The first-order derivative of a constant function vanishes identically. Also the Caputo derivative with order α of a constant function is zero: $\partial_t^\alpha 1 = 0$. However, the Riemann–Liouville derivative is not zero. More precisely,

$$D_t^\alpha 1 = \frac{t^{-\alpha}}{\Gamma(1-\alpha)}.$$

The fundamental theorem of calculus reads

$$\int_0^t \frac{du}{ds}(s) ds = u(t) - u(0), \quad 0 < t < T,$$

which indicates that the integration is a converse operation to the differentiation. We can have a similar inversion for d_t^α and D_t^α, whose special case is (1.2).

More precisely, as an important formula, we can prove

$$J^\alpha d_t^\alpha u(t) = u(t), \quad 0 < t < T \tag{1.9}$$

for $u \in C[0, T] \cap C^1(0, T]$ satisfying

$$\int_0^T \left| \frac{du}{dt}(t) \right| dt < \infty \tag{1.10}$$

and

$$u(0) = 0. \tag{1.11}$$

Moreover,

$$J^\alpha D_t^\alpha u(t) = u(t), \quad 0 < t < T \tag{1.12}$$

for $u \in C[0, T] \cap C^1(0, T]$ satisfying (1.10).

Proof of (1.9) First we obtain

$$J^\alpha d_t^\alpha u(t) = J^\alpha \left(\frac{1}{\Gamma(1-\alpha)} \int_0^s (s-\xi)^{-\alpha} \frac{du}{d\xi}(\xi) d\xi \right)$$

$$= \frac{1}{\Gamma(\alpha)\Gamma(1-\alpha)} \int_0^t (t-s)^{\alpha-1} \left(\int_0^s (s-\xi)^{-\alpha} \frac{du}{d\xi}(\xi) d\xi \right) ds$$

$$= \frac{1}{\Gamma(\alpha)\Gamma(1-\alpha)} \int_0^t \frac{du}{d\xi}(\xi) \left(\int_\xi^t (t-s)^{\alpha-1}(s-\xi)^{-\alpha} ds \right) d\xi.$$

For the last equality, we used the exchange of orders of integrals: $\int_0^t \left(\int_0^s \cdots d\xi \right) ds = \int_0^t \left(\int_\xi^t \cdots ds \right) d\xi$. By (1.7) we conclude that the right-hand side is $\int_0^t \frac{du}{d\xi}(\xi) d\xi = u(t) - u(0)$, which completes the proof of (1.9). The proof of (1.12) is similar and so is omitted. ∎

For the Caputo derivative, the usual formulae for derivatives of orders of natural numbers do not hold. For example, for $u, v \in C^1[0, T]$, we have

$$d_t^\alpha (u + v)(t) = d_t^\alpha u(t) + d_t^\alpha v(t),$$

$$d_t^\alpha (cu)(t) = c d_t^\alpha u(t), \quad \text{where } c \text{ is a constant,}$$

but

$$d_t^\alpha(uv)(t) \neq (d_t^\alpha u(t))v(t) + u(t)(d_t^\alpha v)(t) \tag{1.13}$$

and

$$(d_t^\alpha(d_t^\beta u))(t) \neq d_t^{\alpha+\beta}u(t). \tag{1.14}$$

The violation of the Leibniz rules (see (1.13)) implies the lack of integration by parts, which causes many difficulties in considering fractional differential equations.

As for (1.14), we consider $u(t) = t^\alpha$ and $\beta = \alpha \in (0,1)$. Then (1.7) yields $d_t^{2\alpha}t^\alpha = \frac{\Gamma(1+\alpha)}{\Gamma(1-\alpha)}t^{-\alpha}$, but $d_t^\alpha(d_t^\alpha t^\alpha) = d_t^\alpha(\Gamma(\alpha+1)) = 0$, which means that $d_t^{2\alpha}t^\alpha \neq d_t^\alpha(d_t^\alpha t^\alpha)$.

As for the successive derivatives, we can directly prove the following. Let $0 < \alpha, \beta < 1$ and $\alpha + \beta < 1$. Then

$$(d_t^\alpha(d_t^\beta u))(t) = d_t^{\alpha+\beta}u(t), \quad 0 \leq t \leq T$$

for $u \in C^1[0, T]$ satisfying $u(0) = (d_t^\beta u)(0) = 0$, and

$$(D_t^\alpha(D_t^\beta u))(t) = (D_t^{\alpha+\beta}u)(t)$$

for $u \in C^1[0, T]$.

It is important that the derivative $\dfrac{du}{dt}$ characterizes the behavior, that is, increase or decrease, of $u(t)$. We can accept that such a property holds in a limited sense for the fractional derivatives, because the fractional derivative at t involves the integral over $(0, t)$. On the other hand, Luchko [20] proves the following extremum principle.

Lemma 1.2 *Let $u \in C[0, T] \cap C^1(0, T]$ satisfy*

$$\int_0^T \left| \frac{du}{dt}(t) \right| dt < \infty. \tag{1.15}$$

We assume that $u(t)$ attains the minimum at some t_1 satisfying $0 < t_1 \leq T: u(t_1) = \min_{0 \leq t \leq T} u(t)$. Then

$$d_t^\alpha u(t_1) \leq 0.$$

Proof We set $y(s) = u(s) - u(t_1)$, $0 \leq s \leq T$. Then $y(s) \geq 0$, $0 \leq s \leq T$ and $d_t^\alpha u(s) = d_t^\alpha y(s)$, $0 \leq s \leq T$. For arbitrary $\varepsilon > 0$, we can choose small $\delta > 0$ such that $\delta < t_1$ and, since $y(s)$ satisfies (1.15),

$$\left| \frac{1}{\Gamma(1-\alpha)} \int_0^\delta (t_1 - s)^\alpha \frac{dy}{ds}(s)ds \right| \leq \frac{1}{\Gamma(1-\alpha)(t_1-\delta)^{-\alpha}} \int_0^\delta \left| \frac{dy}{ds}(s) \right| ds < \varepsilon.$$

In defining $d_t^\alpha y(t_1)$, we divide the integration interval into $(0, \delta)$ and (δ, t_1), so that for already chosen δ, we have

$$d_t^\alpha y(t_1) = \frac{1}{\Gamma(1-\alpha)} \int_0^\delta (t_1 - s)^{-\alpha} \frac{dy}{ds}(s)ds + \frac{1}{\Gamma(1-\alpha)} \int_\delta^{t_1} (t_1 - s)^{-\alpha} \frac{dy}{ds}(s)ds =: I_1 + I_2.$$

$$(1.16)$$

Then $|I_1| < \varepsilon$. Moreover, for $\delta > 0$, since $y \in C^1(0, T]$ and $y(t_1) = 0$, we can choose a constant $C_\delta > 0$ such that $|y(s)| \leq C_\delta |t_1 - s|$, $\delta \leq s \leq T$. Therefore integrating by parts, we have

$$I_2 = -\frac{(t_1 - \delta)^{-\alpha} y(\delta)}{\Gamma(1-\alpha)} - \frac{\alpha}{\Gamma(1-\alpha)} \int_\delta^{t_1} (t_1 - s)^{-\alpha-1} y(s)ds.$$

By $t_1 - \delta > 0$ and $y(s) \geq 0$, $0 \leq s \leq T$, we see that $I_2 \leq 0$. Hence by (1.16), we obtain $\partial_t^\alpha y(t_1) \leq \varepsilon$. Here since $\varepsilon > 0$ is arbitrary, we let $\varepsilon \to 0$ and the proof of Lemma 1.2 is complete. ∎

Lemma 1.2 does not hold for $\alpha > 1$. It follows from Lemma 1.2 that if $d_t^\alpha u(t) > 0$, $0 \leq t \leq T$, then $u(t) \geq u(0)$ for all $t \in [0, T]$.

In fact, assume that there exists $t_0 \in [0, T]$ such that $u(t_0) = \min_{0 \leq t \leq T} u(t) < u(0)$. Then $t_0 \neq 0$. That is, $0 < t_0 \leq T$. Lemma 1.2 yields that $d_t^\alpha u(t_0) \leq 0$, which contradicts the assumption $d_t^\alpha u(t) > 0$. Thus the proof is completed.

This is much weaker than the usual property for $\frac{du}{dt}$, and we do not know whether $u(t_0) \geq u(t_1)$ if $d_t^\alpha u(t) \leq 0$, $t_0 < t < t_1$ with $0 < t_0 < t_1 < T$.

Throughout this book, we set

$$\begin{cases} W^{1,1}(0, T) := \left\{ u \in L^1(0, T); \frac{du}{dt} \in L^1(0, T) \right\}, \\ {}_0W^{1,1}(0, T) := \{ u \in W^{1,1}(0, T); \ u(0) = 0 \} \end{cases} \tag{1.17}$$

with the norm

$$\|u\|_{W^{1,1}(0,T)} = \|u\|_{L^1(0,T)} + \left\| \frac{du}{dt} \right\|_{L^1(0,T)}.$$

We note that $u \in W^{1,1}(0, T)$ if and only if there exists an absolutely continuous function $\tilde{u}$ on $[0, T]$ such that $\tilde{u}(t) = u(t)$ for almost all $t \in (0, T)$. Here and henceforth we identify $\tilde{u}$ with u for $u \in W^{1,1}(0, T)$.

We close this chapter with the following lemma which generalizes Lemma 1.1 to $u \in W^{1,1}(0, T)$,

Lemma 1.3

(i) For $u \in W^{1,1}(0, T)$, we have $J^{1-\alpha} u \in W^{1,1}(0, T)$ and

$$\frac{d}{dt} J^{1-\alpha} u(t) = J^{1-\alpha} \frac{du}{dt}(t) + \frac{u(0)}{\Gamma(1-\alpha)} t^{-\alpha}, \quad 0 < t < T.$$

(ii) Let $u_n, u \in W^{1,1}(0, T)$ and $u_n \longrightarrow u$ in $W^{1,1}(0, T)$. Then $d_t^\alpha u_n \longrightarrow d_t^\alpha u$ in $L^1(0, T)$.

(iii) $d_t^\alpha u = D_t^\alpha u$ for $u \in W^{1,1}(0, T)$ satisfying $u(0) = 0$.

(iv) Let $\alpha, \beta > 0$. Then

$$J^\alpha(J^\beta u) = J^{\alpha+\beta}u, \quad u \in L^2(0, T).$$

By the Sobolev embedding (e.g., formula (1.20) on p. 193 in Kato [13], Theorem 5.4 in Adams [2]), we see that $W^{1,1}(0, T) \subset C[0, T]$ and so $u(0)$ makes sense for $u \in W^{1,1}(0, T)$. We can interpret Lemma 1.3 (i) so that the exchange of $\frac{d}{dt}$ and $\int_0^t \cdots ds$ is possible in

$$\int_0^t (t - s)^{-\alpha} u(s) ds.$$

Proof (i) and (iii) By $u \in W^{1,1}(0, T)$, we have

$$u(s) = \int_0^s \frac{du}{d\xi}(\xi) d\xi + u(0), \quad 0 < s < T.$$

Hence

$$(J^{1-\alpha}u)(t) = \frac{1}{\Gamma(1-\alpha)} \int_0^t (t-s)^{-\alpha} \left(\int_0^s \frac{du}{d\xi}(\xi) d\xi + u(0) \right) ds$$

$$= \frac{1}{\Gamma(1-\alpha)} \int_0^t (t-s)^{-\alpha} \left(\int_0^s \frac{du}{d\xi}(\xi) d\xi \right) ds + \frac{1}{\Gamma(1-\alpha)} \left(\int_0^t (t-s)^{-\alpha} ds \right) u(0)$$

$$= \frac{1}{\Gamma(1-\alpha)} \int_0^t \left(\int_\xi^t (t-s)^{-\alpha} ds \right) \frac{du}{d\xi}(\xi) d\xi + \frac{1}{\Gamma(1-\alpha)} \frac{t^{1-\alpha}}{1-\alpha} u(0).$$

Here we have exchanged the orders of integrals. Therefore

$$(J^{1-\alpha}u)(t) = \frac{1}{(1-\alpha)\Gamma(1-\alpha)} \int_0^t (t-\xi)^{1-\alpha} \frac{du}{d\xi}(\xi) d\xi$$

$$+ \frac{1}{(1-\alpha)\Gamma(1-\alpha)} t^{1-\alpha} u(0), \quad 0 < t < T.$$

Taking the differentiation in t, noting that $1 - \alpha > 0$, we see that $J^{1-\alpha}u \in W^{1,1}(0, T)$ and

$$\frac{d}{dt} J^{1-\alpha} u(t) = J^{1-\alpha} \frac{du}{dt}(t) + \frac{u(0)}{\Gamma(1-\alpha)} t^{-\alpha}. \tag{1.18}$$

Here we use Lemma A.1 in the Appendix, which is called the Young inequality for the convolution:

$$J^{1-\alpha}\frac{du}{dt}(t) = \left(\frac{1}{\Gamma(1-\alpha)}t^{-\alpha} * \frac{du}{dt}\right)(t)$$

and

$$\left\|J^{1-\alpha}\frac{du}{dt}\right\|_{L^1(0,T)} \le \left\|\frac{1}{\Gamma(1-\alpha)}t^{-\alpha}\right\|_{L^1(0,T)} \left\|\frac{du}{dt}\right\|_{L^1(0,T)} < \infty.$$

Moreover, (1.18) directly implies conclusion (iii).   ■

Proof of (ii) We have

$$d_t^\alpha u_n(t) - d_t^\alpha u(t) = \frac{1}{\Gamma(1-\alpha)}\int_0^t (t-s)^{-\alpha}\left(\frac{du_n}{ds}(s) - \frac{du}{ds}(s)\right) ds$$

$$= \left(\frac{1}{\Gamma(1-\alpha)}t^{-\alpha} * \frac{d}{dt}(u_n - u)\right)(t).$$

Again the Young inequality Lemma A.1 for the convolution yields

$$\|d_t^\alpha u_n - d_t^\alpha u\|_{L^1(0,T)} \le \frac{1}{\Gamma(1-\alpha)}\|t^{-\alpha}\|_{L^1(0,T)} \left\|\frac{d}{dt}(u_n - u)\right\|_{L^1(0,T)}$$

$$\le \frac{1}{\Gamma(1-\alpha)}\frac{T^{1-\alpha}}{1-\alpha}\|u_n - u\|_{W^{1,1}(0,T)}.$$

Thus the proof of Lemma 1.3 (ii) is complete.   ■

Proof of (iv) Exchanging the orders of the integrals and applying (1.7), we verify

$$J^\alpha J^\beta u(t) = \frac{1}{\Gamma(\alpha)}\frac{1}{\Gamma(\beta)}\int_0^t (t-s)^{\alpha-1}\left(\int_0^s (s-\xi)^{\beta-1}u(\xi)d\xi\right) ds$$

$$= \frac{1}{\Gamma(\alpha)\Gamma(\beta)}\int_0^t \left(\int_\xi^t (t-s)^{\alpha-1}(s-\xi)^{\beta-1}ds\right) u(\xi)d\xi$$

$$= \frac{1}{\Gamma(\alpha)\Gamma(\beta)}\int_0^t \frac{\Gamma(\alpha)\Gamma(\beta)}{\Gamma(\alpha+\beta)}(t-s)^{\alpha+\beta-1}u(\xi)d\xi = J^{\alpha+\beta}u(t).$$

■

Chapter 2
Definition of Fractional Derivatives in Sobolev Spaces and Properties

2.1 Motivations

We consider a very simple equation

$$d_t^\alpha u(t) = f(t), \quad 0 < t < T, \qquad u(0) = a \tag{2.1}$$

and find $u(t)$ satisfying (2.1) with given $f \in L^2(0, T)$ and $a \in \mathbb{R}$. Needless to say, for $\alpha = 1$, we have

$$u(t) = \int_0^t f(s)ds + a, \quad 0 < t < T.$$

If $\alpha = 1$ and u satisfies (2.1) with given $f \in L^1(0, T)$, then $u \in C[0, T]$, so that $u(0)$ can be calculated by substituting $t = 0$ into $u(t)$. In other words, for $\alpha = 1$, for any $f \in L^2(0, T)$ and $a \in \mathbb{R}$, there exists a unique $u \in H^1(0, T)$ satisfying (2.1). The situation is different from the case of $0 < \alpha < 1$. In most of the existing works, for $f \in L^2(0, T)$ or $f \in L^1(0, T)$, detailed regularity of the solution to $d_t^\alpha u = f$ is not clarified and so the initial condition can only be justified if the solutions are proved to possess a certain regularity at $t = 0$. The class $f \in L^2(0, T)$ is not sufficient for such regularity. For example, Theorem 3.3 (p. 126) in Podlubny [24] assumes that $f \in C[0, T]$.

For fractional differential equations, the Laplace transform:

$$(Lu)(p) := \int_0^\infty e^{-pt} u(t)dt, \quad p > p_0 : \quad \text{some constant}$$

© The Author(s), under exclusive licence to Springer Nature Singapore Pte Ltd. 2020
A. Kubica et al., *Time-Fractional Differential Equations*, SpringerBriefs in Mathematics, https://doi.org/10.1007/978-981-15-9066-5_2

is convenient, but the justification of the initial value is indispensable in order to use the Laplace transform. Because by some repetition of calculations, the formula

$$L(d_t^\alpha u)(p) = p^\alpha (Lu)(p) - p^{\alpha-1} u(0)$$

is known (e.g., p. 106 in [24]) and we have to justify the sense of $u(0)$, which requires a certain smoothness of u at $t = 0$. Such regularity at $t = 0$ is not well established for $f \in L^2(0, T)$. Moreover, for such f, we have to justify the solution formula for (2.1) (e.g., Kilbas et al. [15, p. 141]).

We illustrate with one simple example. Let $0 < \alpha < \frac{1}{2}$ and

$$f(t) = t^{\delta-\frac{1}{2}}, \quad 0 < t < T,$$

where $\delta > 0$ is a constant. We consider

$$d_t^\alpha u(t) = t^{\delta-\frac{1}{2}}, \quad u(0) = a. \tag{2.2}$$

Since $\delta - \frac{1}{2} > -1$, we can formally apply the solution formula (e.g., [15, p. 141]) and obtain

$$u(t) = a + \frac{1}{\Gamma(\alpha)} \int_0^t (t-s)^{\alpha-1} s^{\delta-\frac{1}{2}} ds = a + C_0 t^{\alpha+\delta-\frac{1}{2}}, \tag{2.3}$$

where we set

$$C_0 = \frac{\Gamma\left(\delta + \frac{1}{2}\right)}{\Gamma\left(\alpha + \delta + \frac{1}{2}\right)}.$$

Moreover, $u(t)$ given by (2.3) cannot satisfy (2.2) if $0 < \alpha < \frac{1}{2}$ and $\delta > 0$ is small such that $\alpha + \delta - \frac{1}{2} < 0$. Indeed, $\lim_{t \downarrow 0} u(t) = \infty$, and so the initial condition does not follow common sense. Furthermore we formally calculate $d_t^\alpha t^{\alpha+\delta-\frac{1}{2}}$:

$$d_t^\alpha t^{\alpha+\delta-\frac{1}{2}} = \frac{1}{\Gamma(1-\alpha)} \int_0^t (t-s)^{-\alpha} \frac{d}{ds}(s^{\alpha+\delta-\frac{1}{2}}) ds = \frac{\alpha+\delta-\frac{1}{2}}{\Gamma(1-\alpha)} \int_0^t (t-s)^{-\alpha} s^{\alpha+\delta-\frac{3}{2}} ds.$$

However, since $\alpha + \delta - \frac{3}{2} < -1$, the integral does not exist. This means that formula (2.3) does not hold for $f \in L^2(0, T)$ in general, although (2.3) is convergent for e.g., $f \in L^\infty(0, T)$ as a solution formula to (2.1). We can verify that (2.3) gives a solution to (2.1) if $\frac{du}{dt} \in L^2(0, T)$, because we can calculate $d_t^\alpha u$ pointwise. In terms of an extended Caputo derivative, we will further discuss this later (see, e.g., (2.36)).

In some references (e.g., Podlubny [24, pp. 78–79]), it is emphasized that the Caputo derivative d_t^α can admit the initial value problem (2.1), unlike the Riemann–Liouville derivative D_t^α. However, this simple example $f(t) = t^{\delta-\frac{1}{2}}$ shows that the pointwise Caputo derivative d_t^α cannot make the initial value problem (2.1) well-posed for general $f \in L^2(0, T)$.

Moreover, for guaranteeing the uniqueness of u satisfying $d_t^\alpha u = f \in L^2(0, T)$, we certainly need some extra conditions. In fact, since $d_t^\alpha 1 = 0$ by the definition, if u satisfies $d_t^\alpha u = f$ in $(0, T)$, then $u + C_1$ satisfies the same equation with arbitrary constant C_1. These discussions suggest the necessity for reconsidering the pointwise Caputo derivative d_t^α and redefining the Caputo derivative within the framework of $L^2(0, T)$.

The function space $L^2(0, T)$ is reasonable and convenient as data space. Hence it is natural to formulate the initial value problem and define d_t^α for $f \in L^2(0, T)$ in order to establish a more unified theory for fractional differential equations. Thus we construct the theory where the fractional derivatives should be included in $L^2(0, T)$. This is our main motivation in this book, and we construct a seemingly different fractional derivative ∂_t^α, but we will prove that it is essentially same as the closure operator of the Caputo derivative in $_0C^1[0, T]$ (see Sect. 2.3).

First in Sects. 2.2 and 2.3, we define the generalized fractional derivative ∂_t^α in Sobolev spaces $H_\alpha(0, T)$. Then in Chaps. 3 and 4, in terms of such ∂_t^α, we formulate an initial value problem and prove the well-posedness.

Our formulation is similar to Zacher [30] (see also Kubica and Yamamoto [17]), but more essentially relies on the property of the generalized fractional derivative ∂_t^α defined later in some Sobolev space. These properties are feasible for the applications such as the clarification of the Sobolev regularity of solutions to initial boundary value problems.

2.2 Preliminaries: Operational Structure of J^α

By the direct observation of the definition (1.4) of the pointwise Caputo derivative, we need the first derivative $\frac{du}{ds}(s)$ in order to define the derivative $d_t^\alpha u$ of order $\alpha < 1$. In order to define such an adequate fractional derivative, which is denoted by ∂_t^α, we should fulfill the following:

1. ∂_t^α should be well-defined in a subspace of the Sobolev space of order α;
2. the norm equivalence between $\|\partial_t^\alpha u\|_{L^2(0,T)}$ and some conventional norm of u such as the norm in a Sobolev space.

For them,

- We will interpret J^α as the fractional power of the operator defined by

$$Ju(t) = \int_0^t u(s)ds$$

with the domain $\mathcal{D}(J) = L^2(0, T)$.
- We define ∂_t^α as the inverse to J^α.

These issues are done respectively in this section and Sect. 2.3.

The arguments in this section and a part of Sect. 2.3 are based on Gorenflo and Yamamoto [9], Gorenflo et al. [11].

By $L^2(0, T)$ and $H^\alpha(0, T)$ we mean the usual L^2-space and the fractional Sobolev space on the interval $(0, T)$ (see e.g., [2, Chapter VII]), respectively, and we define the norm in $H^\alpha(0, T)$ by

$$\|u\|_{H^\alpha(0,T)} := \left(\|u\|^2_{L^2(0,T)} + \int_0^T \int_0^T \frac{|u(t) - u(s)|^2}{|t - s|^{1+2\alpha}} dt ds \right)^{\frac{1}{2}}.$$

The L^2-norm and the scalar product in L^2 are denoted by $\| \cdot \|_{L^2(0,T)}$ and $(\cdot, \cdot) = (\cdot, \cdot)_{L^2(0,T)}$, respectively. By $\sim$ we denote a norm equivalence. Since J^α is injective in $L^2(0, T)$, by $J^{-\alpha}$ we denote the algebraic inverse to J^α.

We set

$$_0H^\alpha(0, T) = \{u \in H^\alpha(0, T); \ u(0) = 0\} \quad \text{if } \frac{1}{2} < \alpha \le 1.$$

We further define the Banach spaces

$$H_\alpha(0, T) := \begin{cases} _0H^\alpha(0, T), & \frac{1}{2} < \alpha < 1, \\ \left\{v \in H^{\frac{1}{2}}(0, T); \int_0^T \frac{|v(t)|^2}{t} dt < \infty \right\}, & \alpha = \frac{1}{2}, \\ H^\alpha(0, T), & 0 < \alpha < \frac{1}{2} \end{cases}$$

with the following norm:

$$\|v\|_{H_\alpha(0,T)} = \begin{cases} \|v\|_{H^\alpha(0,T)}, \ 0 < \alpha < 1, \ \alpha \ne \frac{1}{2}, \\ \left(\|v\|^2_{H^{\frac{1}{2}}(0,T)} + \int_0^T \frac{|v(t)|^2}{t} dt\right)^{\frac{1}{2}}, \ \alpha = \frac{1}{2}. \end{cases} \tag{2.4}$$

For later arguments, we need some convenient and not too narrow subspace of $H_\alpha(0, T)$. In particular, for $H_{\frac{1}{2}}(0, T)$, such a subspace is more delicate.

We recall

$$_0W^{1,1}(0, T) = \{u \in W^{1,1}(0, T); \ u(0) = 0\}$$

and introduce the following space:

$$W_\alpha(0, T) := \left\{ u \in W^{1,1}(0, T); \ \text{there exists a constant } C_u > 0 \text{ such that} \right.$$

$$\left. \left| \frac{du}{dt}(t) \right| \le C_u t^{\alpha-1} \text{ almost all } t, \quad u(0) = 0 \right\}. \tag{2.5}$$

Here $C_u > 0$ depends on a choice of u. We note that $t^\beta \in W_\alpha(0, T)$ for any $\beta \ge \alpha$.

Henceforth by $C > 0$, $C_1 > 0$, etc., we denote generic constants which are independent of functions under consideration but dependent on α, T, while C_u means that it depends on a function or a quantity u under consideration.

Then we can prove:

Lemma 2.1 *Let* $0 < \alpha < 1$. *Then*

$$W_\alpha(0, T) \subset H_\alpha(0, T).$$

Proof First we prove

$$\int_0^T \int_0^T (s^{\gamma_1} + t^{\gamma_1})|t - s|^{\gamma_2} ds\, dt < \infty \quad \text{if } \gamma_1, \gamma_2 > -1. \tag{2.6}$$

∎

Proof of (2.6) The conclusion (2.6) is trivial for $\gamma_1 \geq$ or $\gamma_2 \geq 0$. Therefore we can assume that $-1 < \gamma_1, \gamma_2 < 0$. We note

$$\int_0^T \int_0^T (s^{\gamma_1} + t^{\gamma_1})|t - s|^{\gamma_2} ds\, dt = 2 \int_0^T \left(\int_0^t (s^{\gamma_1} + t^{\gamma_1})|t - s|^{\gamma_2} ds \right) dt. \tag{2.7}$$

Indeed,

$$\int_0^T \int_0^T (s^{\gamma_1} + t^{\gamma_1})|t - s|^{\gamma_2} ds\, dt = \int_0^T \left(\left(\int_0^t + \int_t^T \right) (s^{\gamma_1} + t^{\gamma_1})|t - s|^{\gamma_2} ds \right) dt.$$

On the other hand,

$$\int_0^T \left(\int_t^T (s^{\gamma_1} + t^{\gamma_1})|t - s|^{\gamma_2} ds \right) dt = \int_0^T \left(\int_0^s (s^{\gamma_1} + t^{\gamma_1})|t - s|^{\gamma_2} dt \right) ds$$

$$= \int_0^T \left(\int_0^t (t^{\gamma_1} + s^{\gamma_1})|s - t|^{\gamma_2} ds \right) dt.$$

In the last equality, we exchange s and t. Then

$$\int_0^T \left(\int_0^T (s^{\gamma_1} + t^{\gamma_1})|t - s|^{\gamma_2} ds \right) dt = 2 \int_0^T \left(\int_0^t (s^{\gamma_1} + t^{\gamma_1})|t - s|^{\gamma_2} ds \right) dt$$

$$= 2 \int_0^T \left(\int_0^t s^{\gamma_1}(t - s)^{\gamma_2} ds \right) dt + 2 \int_0^T t^{\gamma_1} \left(\int_0^t (t - s)^{\gamma_2} ds \right) dt$$

$$= 2 \frac{\Gamma(\gamma_1 + 1)\Gamma(\gamma_2 + 1)}{\Gamma(\gamma_1 + \gamma_2 + 2)} \int_0^T t^{\gamma_1 + \gamma_2 + 1} dt + 2 \int_0^T \frac{t^{\gamma_1 + \gamma_2 + 1}}{1 + \gamma_2} dt < \infty$$

by $\gamma_1 > -1$ and $\gamma_2 > -1$. Thus the proof of (2.6) is completed. ∎

Now we complete the proof of Lemma 2.1. Indeed, $u(t) = \int_0^t \frac{du}{ds}(s)$ for $u \in W_\alpha(0, T)$. Hence

$$|u(t) - u(s)| = \left| \int_s^t \frac{du}{d\xi}(\xi) d\xi \right| \leq \left| \int_s^t \left| \frac{du}{d\xi}(\xi) \right| d\xi \right|$$

$$\leq C_u \min_{\xi \in (s,t)} |\xi|^{\alpha-1}(t - s) \leq C_u(t^{\alpha-1} + s^{\alpha-1})|t - s|.$$

Let $\frac{1}{2} < \alpha < 1$. Then

$$\frac{|u(t) - u(s)|^2}{|t - s|^{1+2\alpha}} \leq C(t^{2\alpha-2} + s^{2\alpha-2})|t - s|^{1-2\alpha}.$$

Since $2\alpha - 2 > -1$ by $\frac{1}{2} < \alpha < 1$ and $1 - 2\alpha > -1$ by $0 < \alpha < 1$, the inequality (2.6) implies

$$\int_0^T \int_0^T \frac{|u(t) - u(s)|^2}{|t - s|^{1+2\alpha}} ds dt < \infty.$$

Next let $0 < \alpha \leq \frac{1}{2}$. Similarly to (2.7), we can verify

$$\int_0^T \int_0^T \frac{|u(t) - u(s)|^2}{|t - s|^{1+2\alpha}} ds dt = 2 \int_0^T \left(\int_0^t \frac{|u(t) - u(s)|^2}{|t - s|^{1+2\alpha}} ds \right) dt.$$

Since

$$\left| \frac{du}{d\xi}(\xi) \right| \leq C_u \xi^{\alpha-1} \leq C_u s^{\alpha-1}$$

for $\xi \in (s, t)$, we choose $\delta > 0$ such that $0 < \delta < \frac{\alpha}{1-\alpha} \leq 1$. Then, by the mean value theorem we can choose $\eta \in (s, t)$ such that

$$|u(t) - u(s)|^2 = |u(t) - u(s)|^{1-\delta}|u(t) - u(s)|^{1+\delta}$$

$$\leq \left(2\|u\|_{L^\infty(0,T)} \right)^{1-\delta} \left(\left| \frac{du}{dt}(\eta) \right| |t - s| \right)^{1+\delta} \leq C(\xi^{\alpha-1}(t - s))^{1+\delta}$$

$$\leq C s^{(\alpha-1)(1+\delta)}(t - s)^{1+\delta}$$

for $0 < s < t < T$. Consequently, since $\alpha + \alpha\delta - \delta > 0$ and $1 + \delta - 2\alpha > 0$ by $0 < \alpha \leq \frac{1}{2}$ and $0 < \delta < \frac{\alpha}{1-\alpha}$, the inequality (2.6) yields

$$\int_0^T \int_0^t \frac{|u(t) - u(s)|^2}{|t - s|^{1+2\alpha}} ds dt \leq C \int_0^T \left(\int_0^t s^{(\alpha+\alpha\delta-\delta)-1}(t - s)^{(1+\delta-2\alpha)-1} ds \right) dt < \infty.$$

Thus we proved $u \in H^\alpha(0, T)$. Therefore $u \in H_\alpha(0, T) = H^\alpha(0, T)$ for $0 < \alpha < \frac{1}{2}$. Next for $\frac{1}{2} < \alpha < 1$, by $u(0) = 0$, we readily see that $u \in H_\alpha(0, T) = {}_0H^\alpha(0, T)$. Therefore $W_\alpha(0, T) \subset H_\alpha(0, T)$ if $0 < \alpha < 1$ and $\alpha \neq \frac{1}{2}$. Moreover, for $\alpha = \frac{1}{2}$, we have

$$|u(t)| = \left| \int_0^t \frac{du}{ds}(s)ds \right| \leq \int_0^t C_u s^{-\frac{1}{2}} ds = 2C_u t^{\frac{1}{2}},$$

and so $\int_0^T \frac{|u(t)|^2}{t} dt < \infty$, which implies that $H_{\frac{1}{2}}(0, T)$. Thus the proof of Lemma 2.1 is completed. ∎

In fact, the space $W_\alpha(0, T)$ is a convenient subspace of $H_\alpha(0, T)$.

Remark 2.1 For $H_{\frac{1}{2}}(0, T)$, Lions and Magenes [19] use a different notation ${}_0H_0^{\frac{1}{2}}(0, T)$ (Remark 11.5 (p. 68) in [19] vol. I). However, we use $H_{\frac{1}{2}}(0, T)$ as well as $H_\alpha(0, T)$, $0 < \alpha < 1$ throughout this book.

Henceforth we set

$$_0C^1[0, T] = \{\varphi \in C^1[0, T];\ \varphi(0) = 0\}$$

and $\overline{Z}^X$ denotes the closure of Z by the norm of X.

Lemma 2.2

$$\overline{_0C^1[0, T]}^{H_\alpha(0,T)} = H_\alpha(0, T).$$

Proof For $\frac{1}{2} < \alpha \leq 1$, we see that $H_\alpha(0, T) = {}_0H^\alpha(0, T) := \{u \in H^\alpha(0, T);\ u(0) = 0\}$, so that the mollifier (e.g., Adams [2]) yields the conclusion. Let $0 \leq \alpha \leq \frac{1}{2}$. By Lions and Magenes [19], we have

$$H_\alpha(0, T) = [{}_0H^1(0, T), L^2(0, T)]_{1-\alpha}$$

which is the interpolation space. Applying Proposition 6.1 (p. 28) in [19], for $\frac{1}{2} < \gamma < 1$, we see that $[{}_0H^1(0, T), L^2(0, T)]_{1-\gamma}$ is dense in $[{}_0H^1(0, T), L^2(0, T)]_{1-\alpha}$ by $1 - \gamma \leq 1 - \alpha$. Therefore

$$\overline{H_\gamma(0, T)}^{H_\alpha(0,T)} = H_\alpha(0, T)$$

for $\frac{1}{2} < \gamma < 1$. As is already proved, we see that

$$\overline{_0C^1[0, T]}^{H_\gamma(0,T)} = H_\gamma(0, T).$$

Both density properties yield

$$\overline{{}_0C^1[0,T]}^{H_\alpha(0,T)} = H_\alpha(0,T)$$

for $0 \le \alpha \le \frac{1}{2}$. Thus the proof of the lemma is completed.   ∎

For $0 < \alpha < \frac{1}{2}$, the proof of the lemma is more direct by Theorem 11.1 (p. 55) in [19].

Lemma 2.2 is useful, because, thanks to the lemma, in order to prove estimates in $H_\alpha(0,T)$, it usually suffices to prove them for ${}_0C^1[0,T]$.

Remark 2.2 By Theorem 11.1 (p. 55) in [19], we see that

$$\overline{{}_0C^1[0,T]}^{H^\alpha(0,T)} = H^\alpha(0,T), \quad 0 \le \alpha \le \frac{1}{2}. \tag{2.8}$$

We should distinguish $H_\alpha(0,T)$ from $H^\alpha(0,T)$. For $\frac{1}{2} < \alpha \le 1$, the mollifier yields that ${}_0C^1[0,T]$ is dense in $H_\alpha(0,T)$. To sum up,

$$\overline{{}_0C^1[0,T]}^{H^\alpha(0,T)} = \begin{cases} H^\alpha(0,T), \, 0 < \alpha \le \frac{1}{2}, \\ H_\alpha(0,T), \, \frac{1}{2} < \alpha < 1. \end{cases} \tag{2.9}$$

By Theorem 2.1 in Gorenflo et al. [11], we know the following theorem.

Theorem 2.1 *Let* $0 < \alpha < 1$.

(i) $J^\alpha : L^2(0,T) \longrightarrow H_\alpha(0,T)$ *is injective and surjective.*
(ii) *There exists a constant* $C > 0$ *such that*

$$C^{-1}\|J^\alpha u\|_{H_\alpha(0,T)} \le \|u\|_{L^2(0,T)} \le C\|J^\alpha u\|_{H_\alpha(0,T)} \tag{2.10}$$

for all $u \in L^2(0,T)$.

Moreover, we can prove

$$J^\alpha L^2(0,T) \subset H_1(0,T), \quad \alpha \ge 1.$$

However, in this book we limit the range of α to $0 < \alpha \le 1$ and we omit further characterization of $J^\alpha L^2(0,T)$.

In terms of Theorem 2.1, $(J^\alpha)^{-1}$ exists and $J^{-\alpha} = (J^\alpha)^{-1}$. Then:

Theorem 2.2 *There exists a constant* $C > 0$ *such that*

$$C^{-1}\|J^{-\alpha}v\|_{L^2(0,T)} \le \|v\|_{H_\alpha(0,T)} \le C\|J^{-\alpha}v\|_{L^2(0,T)} \tag{2.11}$$

for all $v \in H_\alpha(0,T)$.

By Theorem 2.1, we see:

Corollary 2.1 *Let* $0 < \alpha < 1$. *Then*

$$J^{-\alpha} J^\alpha u = u, \quad u \in L^2(0, T)$$

and

$$J^\alpha J^{-\alpha} u = u, \quad u \in H_\alpha(0, T).$$

The first equality is directly seen by the definition, while the second equality is verified as follows. For $u \in H_\alpha(0, T)$, Theorem 2.1 (i) yields the existence of $w \in L^2(0, T)$ satisfying $u = J^\alpha w$. Therefore $J^\alpha J^{-\alpha} u = J^\alpha J^{-\alpha} (J^\alpha w) = J^\alpha w$. Hence $J^\alpha J^{-\alpha} u = u$ for $u \in H_\alpha(0, T)$.

Henceforth we write for example (2.10)

$$\|J^{-\alpha} v\|_{L^2(0,T)} \sim \|v\|_{H_\alpha(0,T)},$$

when there is no fear of confusion.

Now we prove Theorem 2.1. By $J = J^1$ we denote the integral $(Jy)(t) = \int_0^t y(s)ds$ for $0 \le t \le T$ and by $I : L^2(0, T) \to L^2(0, T)$ the identity mapping. In this section, we consider the space $L^2(0, T)$ over $\mathbb{C}$ with the scalar product $(u, v) = (u, v)_{L^2(0,T)} = \int_0^T u(t)\overline{v(t)}dt$ for complex-valued u, v, and Re η and Im η denote the real and the imaginary parts of a complex number η, respectively. Let $\bar\eta$ denote the complex conjugate of $\eta \in \mathbb{C}$. Henceforth if there is no fear of confusion, then by $(\cdot, \cdot)$ we mean $(\cdot, \cdot)_{L^2(0,T)}$, and $\mathcal{R}(A)$ and $\mathcal{D}(A)$ denote the range and the domain of an operator $A: A\mathcal{D}(A) = \mathcal{R}(A)$ respectively.

Lemma 2.3 *For any* $u \in L^2(0, T)$, *the inequality* Re $(Ju, u) \ge 0$ *holds true and* $\mathcal{R}(I + J) = L^2(0, T)$.

Proof With the notations Re $Ju(t) = \varphi(t)$ and Im $Ju(t) = \psi(t)$, the following chain of equalities and a final estimate can be easily obtained:

$$\text{Re } (Ju, u) = \text{Re } \int_0^T \left(\int_0^t u(s)ds \right) \bar{u}(t)dt = \text{Re } \int_0^T Ju(t)\frac{d}{dt}\overline{Ju(t)}dt$$

$$= \int_0^T \left(\varphi(t)\frac{d\varphi}{dt} + \psi(t)\frac{d\psi}{dt} \right) dt = \frac{1}{2}(\varphi(t)^2 + \psi(t)^2)|_{t=0}^{t=T} = \frac{1}{2}|Ju(T)|^2 \ge 0.$$

Therefore Re $(Ju, u) \ge 0$ for $u \in L^2(0, T)$. Next for $\lambda \ne 0$, we have the representation

$$(\lambda I + J)^{-1}u(t) = \lambda^{-1}u(t) - \lambda^{-2} \int_0^t e^{-(t-s)/\lambda}u(s)ds, \quad 0 \le t \le T \quad (2.12)$$

for $u \in L^2(0, T)$. We will prove (2.12). We set $v(t) = (\lambda I + J)^{-1} u(t)$, that is,

$$\lambda v(t) + \int_0^t v(s)ds = u(t), \quad 0 < t < T. \tag{2.13}$$

Let $u \in C^1[0, T]$. Then

$$v(0) = \frac{1}{\lambda} u(0). \tag{2.14}$$

Differentiating (2.13) with respect to t, we have

$$\frac{dv}{dt}(t) + \frac{1}{\lambda} v(t) = \frac{1}{\lambda} \frac{du}{dt}(t).$$

With (2.14), we obtain

$$v(t) = \frac{1}{\lambda} e^{-\frac{1}{\lambda} t} \left(\int_0^t e^{\frac{1}{\lambda} s} \frac{du}{dt}(s)ds + u(0) \right).$$

Using integration by parts, we have

$$v(t) = \frac{1}{\lambda} u(t) - \frac{1}{\lambda^2} \int_0^t e^{-\frac{t-s}{\lambda}} u(s)ds, \quad 0 < t < T. \tag{2.15}$$

Next let $u \in L^2(0, T)$. Then the right-hand side of (2.15) defines a function $v \in L^2(0, T)$. Exchanging the orders of the integrals: $\int_0^t \left(\int_0^s d\xi \right) ds = \int_0^t \left(\int_\xi^t ds \right) d\xi$, we calculate

$$\int_0^t v(s)ds = \frac{1}{\lambda} \int_0^t u(s)ds - \frac{1}{\lambda^2} \int_0^t \left(\int_0^s e^{-\frac{s-\xi}{\lambda}} u(\xi)d\xi \right) ds$$

$$= \frac{1}{\lambda} \int_0^t u(s)ds - \frac{1}{\lambda^2} \int_0^t \left(\int_\xi^t e^{-\frac{s}{\lambda}} ds \right) e^{\frac{\xi}{\lambda}} u(\xi)d\xi$$

$$= \frac{1}{\lambda} \int_0^t u(s)ds + \frac{1}{\lambda} \int_0^t \left[e^{-\frac{s}{\lambda}} \right]_{s=\xi}^{s=t} e^{\frac{\xi}{\lambda}} u(\xi)d\xi$$

$$= \frac{1}{\lambda} \int_0^t u(s)ds + \frac{1}{\lambda} e^{-\frac{t}{\lambda}} \int_0^t e^{\frac{\xi}{\lambda}} u(\xi)d\xi - \frac{1}{\lambda} \int_0^t u(\xi)d\xi$$

$$= \frac{1}{\lambda} e^{-\frac{t}{\lambda}} \int_0^t e^{\frac{s}{\lambda}} u(s)ds.$$

This and (2.15) yield $\lambda v(t) + \int_0^t v(s)ds = u(t), 0 < t < T$ for each $u \in L^2(0, T)$, which is (2.12).

Setting $\lambda = 1$, by (2.12) the operator $(I + J)^{-1}u$ is defined for all $u \in L^2(0, T)$, which implies that $\mathcal{R}(I + J) = L^2(0, T)$. The proof of Lemma 2.3 is completed. ∎

In terms of Lemma 2.3, for $0 < \alpha < 1$, we can define the fractional power of J, which we denote by $J(\alpha)$ (e.g., Tanabe [27, Chapter 2]). More precisely, we have the formula

$$J(\alpha)u = \frac{\sin \pi \alpha}{\pi} \int_0^\infty \lambda^{\alpha-1}(\lambda I + J)^{-1} Ju \, d\lambda, \quad u \in \mathcal{D}(J) = L^2(0, T) \quad (2.16)$$

(see also Chapter 2, §3 in [27]).

Next we will prove that the fractional power $J(\alpha)$ of the integral operator J coincides with the Riemann–Liouville fractional integral operator J^α on $L^2(0, T)$:

Lemma 2.4

$$(J(\alpha)u)(t) = (J^\alpha u)(t), \quad 0 \le t \le T, \ u \in L^2(0, T), \ 0 < \alpha < 1.$$

Proof By (2.12), we have

$$(\lambda I + J)^{-1} Ju(t) = \lambda^{-1} \int_0^t e^{-(t-s)/\lambda} u(s) ds, \quad u \in L^2(0, T)$$

and by change of variables $\eta = \frac{t-s}{\lambda}$, we obtain

$$\frac{\sin \pi \alpha}{\pi} \int_0^\infty \lambda^{\alpha-1}(\lambda I + J)^{-1} Ju(t) \, d\lambda$$

$$= \frac{\sin \pi \alpha}{\pi} \int_0^\infty \lambda^{\alpha-2} \left(\int_0^t e^{-(t-s)/\lambda} u(s) ds \right) d\lambda$$

$$= \frac{\sin \pi \alpha}{\pi} \int_0^t u(s) \left(\int_0^\infty \lambda^{\alpha-2} e^{-(t-s)/\lambda} d\lambda \right) ds$$

$$= \frac{\sin \pi \alpha}{\pi} \int_0^t u(s) \left(\int_0^\infty \eta^{-\alpha} e^{-\eta} d\eta \right) (t-s)^{\alpha-1} ds$$

$$= \frac{\Gamma(1-\alpha) \sin \pi \alpha}{\pi} \int_0^t u(s)(t-s)^{\alpha-1} ds.$$

Here we use also

$$\int_0^\infty x^\gamma e^{-ax} dx = \frac{\Gamma(\gamma+1)}{a^{\gamma+1}}, \quad a > 0, \ \gamma > -1. \quad (2.17)$$

Now the known formula $\Gamma(1-\alpha)\Gamma(\alpha) = \frac{\pi}{\sin \pi \alpha}$ implies the statement of the lemma.

Next we consider the differential operator

$$
\begin{cases}
(Su)(t) = -\dfrac{d^2 u(t)}{dt^2}, \quad 0 < t < T, \\
\mathcal{D}(S) = \left\{ u \in H^2(0, T);\ u(0) = \dfrac{du}{dt}(T) = 0 \right\}.
\end{cases}
\tag{2.18}
$$

Here we note that the boundary conditions $u(0) = \frac{du}{dt}(T) = 0$ should be interpreted as the traces of u in the Sobolev space $H^2(0, T)$ (see e.g., [2, 19]). It is possible to define the fractional power $S^{\frac{\alpha}{2}}$ of the differential operator S for $0 \le \alpha \le 1$ in terms of the eigenvalues and the eigenfunctions of the eigenvalue problem for the operator S. More precisely, let $0 < \lambda_1 < \lambda_2 < \cdots$ be the eigenvalues and $\psi_k, k \in \mathbb{N}$ the well known corresponding normed eigenfunctions of S. It is easy to derive the explicit formulas for $\lambda_k,\ \psi_k,\ k \in \mathbb{N}$, namely, $\lambda_k = \frac{(2k-1)^2 \pi^2}{4T^2}$ and $\psi_k(t) = \frac{\sqrt{2}}{\sqrt{T}} \sin \sqrt{\lambda_k} t$. In particular, we note that $\psi_k(0) = 0$ and $\psi_k \in H^2(0, T)$. It is known that $\{\psi_k\}_{k \in \mathbb{N}}$ is an orthonormal basis of $L^2(0, T)$. Then the fractional power $S^{\frac{\alpha}{2}}$, $0 \le \alpha \le 1$ of the differential operator S is defined by the relations

$$
\begin{cases}
S^{\frac{\alpha}{2}} u = \sum_{k=1}^{\infty} \lambda_k^{\frac{\alpha}{2}} (u, \psi_k)_{L^2(0,T)} \psi_k, \quad u \in \mathcal{D}(S^{\frac{\alpha}{2}}), \\
\mathcal{D}(S^{\frac{\alpha}{2}}) = \{ u \in L^2;\ \sum_{k=1}^{\infty} \lambda_k^{\alpha} |(u, \psi_k)_{L^2(0,T)}|^2 < \infty \}, \\
\|u\|_{\mathcal{D}(S^{\frac{\alpha}{2}})} = \left(\sum_{k=1}^{\infty} \lambda_k^{\alpha} |(u, \psi_k)_{L^2(0,T)}|^2 \right)^{\frac{1}{2}}.
\end{cases}
\tag{2.19}
$$

According to [9], the domain $\mathcal{D}(S^{\frac{\alpha}{2}})$ can be described as follows:

$$
\mathcal{D}(S^{\frac{\alpha}{2}}) = H_\alpha(0, T).
\tag{2.20}
$$

The relation (2.20) holds not only algebraically but also topologically:

$$
\|S^{\frac{\alpha}{2}} v\|_{L^2(0,T)} \sim \|v\|_{H_\alpha(0,T)}, \quad 0 \le \alpha \le 1,\ v \in \mathcal{D}(S^{\frac{\alpha}{2}}).
\tag{2.21}
$$

In particular, the inclusion $\mathcal{D}(S^{\frac{\alpha}{2}}) \subset H^\alpha(0, T)$ holds true.

Now we are ready to prove Theorem 2.1.   ■

Proof of Theorem 2.1 We first state the Heinz–Kato inequality (e.g., Theorem 2.3.4 in Tanabe [27]): let X be a Hilbert space and linear operators A, B in X satisfy $\mathcal{D}(A) = \mathcal{D}(B)$, $\mathrm{Re}\,(Au, u) \ge 0$, $\mathrm{Re}\,(Bu, u) \ge 0$ for $u \in \mathcal{D}(A)$ and $\mathcal{R}(I + A) = \mathcal{R}(I + B) = X$. We assume that there exists a constant $C > 0$ such that

$$
\|Bu\| \le C\|Au\|, \quad u \in \mathcal{D}(A).
$$

Then, for $0 \le \alpha \le 1$, we see that $\mathcal{D}(A^\alpha) = \mathcal{D}(B^\alpha)$ and

$$
\|B^\alpha u\| \le e^{\pi \sqrt{\alpha(1-\alpha)}} C^\alpha \|A^\alpha u\|, \quad u \in \mathcal{D}(A^\alpha).
$$

The proof of the theorem is done by using the norm equivalence between $\| \cdot \|_{\mathcal{D}(S^{\frac{\alpha}{2}})}$ and $H_\alpha(0, T)$, which is justified by the Heinz–Kato inequality.

First of all, it can be directly verified that $\mathcal{D}(J^{-1}) = J(L^2(0, T)) = {}_0H^1(0, T)$, $(J^{-1}w)(t) = \dfrac{dw(t)}{dt}$, and $\|J^{-1}v\|_{L^2(0,T)} = \|v\|_{H^1(0,T)}$ for $v \in {}_0H^1(0, T)$.

Therefore by (2.21) we obtain the norm equivalence

$$\|J^{-1}v\|_{L^2(0,T)} \sim \|S^{\frac{1}{2}}v\|_{L^2(0,T)}, \quad v \in {}_0H^1(0, T) = \mathcal{D}(J^{-1}) = \mathcal{D}(S^{\frac{1}{2}}).$$

Direct calculations show that

$$\int_0^T \frac{d(\varphi + i\psi)}{dt}(\varphi - i\psi)dt \geq 0,$$

and so both J^{-1} and $S^{\frac{1}{2}}$ satisfy the conditions for the Heinz–Kato inequality. Hence the Heinz–Kato inequality yields

$$\|J^{-\alpha}v\|_{L^2(0,T)} \sim \|S^{\frac{\alpha}{2}}v\|_{L^2(0,T)}, \quad v \in \mathcal{D}(S^{\frac{\alpha}{2}}), \quad \mathcal{D}(J^{-\alpha}) = \mathcal{D}(S^{\frac{\alpha}{2}}). \tag{2.22}$$

By (2.20) and (2.22), we have $H_\alpha(0, T) = \mathcal{D}(S^{\frac{\alpha}{2}}) = \mathcal{D}(J^{-\alpha}) = \mathcal{R}(J^\alpha)$, so that Theorem 2.1 (i) follows. By (2.21) and (2.22), the norm equivalence $\|J^{-\alpha}v\|_{L^2(0,T)} \sim \|v\|_{H_\alpha(0,T)}$ holds true for $v \in \mathcal{D}(J^{-\alpha}) = \mathcal{R}(J^\alpha)$. Next, setting $v = J^\alpha u \in \mathcal{D}(J^{-\alpha})$ with any $u \in L^2(0, T)$, by (2.21), we obtain the following norm equivalence:

$$\|u\|_{L^2(0,T)} \sim \|S^{\frac{\alpha}{2}}(J^\alpha u)\|_{L^2(0,T)} \sim \|J^\alpha u\|_{H^\alpha(0,T)}, \quad u \in L^2(0, T).$$

Therefore Theorem 2.1 (ii) follows. ∎

2.3 Definition of Generalized ∂_t^α in $H_\alpha(0, T)$

Next we show a generalization of Theorem 2.1 (ii).

Theorem 2.3 Let $0 \leq \alpha \leq 1$ and $0 \leq \beta \leq 1$.

(i) Let $0 < \alpha + \beta \leq 1$. Then $J^\alpha : H_\beta(0, T) \longrightarrow H_{\alpha+\beta}(0, T)$ is surjective and

$$\|J^\alpha u\|_{H_{\alpha+\beta}(0,T)} \sim \|u\|_{H_\beta(0,T)}, \quad u \in H_\beta(0, T). \tag{2.23}$$

(ii) Let $0 \leq \beta - \alpha \leq 1$. Then

$$J^{\alpha-\beta}u = J^{-\beta}J^\alpha u, \quad u \in H_{\beta-\alpha}(0, T). \tag{2.24}$$

Remark 2.3 Since J^α is defined by the fractional power of the operator J and $J^{-\alpha}$ by $J^{-\alpha} = (J^\alpha)^{-1}$ for $0 \le \alpha \le 1$, the general theory (e.g., Tanabe [27]) implies

$$J^\alpha(J^\beta u) = J^{\alpha+\beta}u, \quad u \in L^2(0, T),$$

provided that $\alpha, \beta, \alpha + \beta \in [-1, 1]$ and we understand

$$\mathcal{D}(J^{-\gamma}) = \begin{cases} L^2(0, T), & \gamma \le 0, \\ H_\gamma(0, T), & \gamma \ge 0. \end{cases}$$

Proof

(i) We note that

$$H_\alpha(0, T) = \{J^\alpha\varphi; \ \varphi \in L^2(0, T)\}, \quad 0 \le \alpha \le 1.$$

First let $u \in J^\alpha(H_\beta(0, T))$, that is, $u = J^\alpha\varphi$ with some $\varphi \in H_\beta(0, T)$. By Theorem 2.1 (i), there exists $\psi \in L^2(0, T)$ such that $\varphi = J^\beta\psi$. Then, noting that J^α is defined as fractional power of J, we see that $u = J^\alpha\varphi = J^\alpha(J^\beta\psi) = J^{\alpha+\beta}\psi$, that is, $u \in H_{\alpha+\beta}(0, T)$ by Theorem 2.1 (i). Hence $J^\alpha(H_\beta(0, T)) \subset H_{\alpha+\beta}(0, T)$. Next let $u \in H_{\alpha+\beta}(0, T)$. Then Theorem 2.1 yields the existence of $\varphi \in L^2(0, T)$ such that $u = J^{\alpha+\beta}\varphi$. Since $J^{\alpha+\beta} = J^\alpha J^\beta$, we obtain $u = J^\alpha(J^\beta\varphi) \in J^\alpha(H_\beta(0, T))$. Hence $J^\alpha : H_\beta(0, T) \longrightarrow H_{\alpha+\beta}(0, T)$ is surjective.

The norm equivalence (2.23) is verified as follows:

$$\|J^\alpha u\|_{H^{\alpha+\beta}(0,T)} \sim \|J^{-\alpha-\beta}J^\alpha u\|_{L^2(0,T)} = \|J^{-\beta}u\|_{L^2(0,T)} \sim \|u\|_{H_\beta(0,T)},$$

by $\|u\|_{H_\gamma(0,T)} \sim \|J^{-\gamma}u\|_{L^2(0,T)}$ and Theorem 2.1 (ii).

(ii) Let $u \in H_{\beta-\alpha}(0, T)$. Replacing β by $\beta - \alpha$ in Theorem 2.3 (i) and applying it, we see that $J^\alpha u \in H_\beta(0, T) = \mathcal{D}(J^{-\beta})$, and so $u \in \mathcal{D}(J^{-\beta}J^\alpha)$. Conversely, let $u \in \mathcal{D}(J^{-\beta}J^\alpha)$. Then $J^\alpha u \in H_\beta(0, T)$. Again applying Theorem 2.3 (i), where we replace β by $\beta - \alpha$, we have $u \in H_{\beta-\alpha}(0, T)$, which implies $H_{\beta-\alpha}(0, T) = \mathcal{D}(J^{-\beta}J^\alpha)$. Since $\mathcal{D}(J^{\alpha-\beta}) = H_{\beta-\alpha}(0, T)$ by the definition of $J^{\alpha-\beta} = (J^{\beta-\alpha})^{-1}$, we obtain $\mathcal{D}(J^{-\beta}J^\alpha) = \mathcal{D}(J^{\alpha-\beta})$. For $u \in H_{\beta-\alpha}(0, T)$, we set $v := J^{\alpha-\beta}u \in L^2(0, T)$. Then $u = J^{\beta-\alpha}v \in H_{\beta-\alpha}(0, T)$. Therefore

$$J^{-\beta}J^\alpha u = J^{-\beta}J^\alpha(J^{\beta-\alpha}v) = J^{-\beta}J^\beta v.$$

Here we used $J^\alpha J^{\beta-\alpha}v = J^\beta v$ by Theorem 2.3 (i). Hence $J^{-\beta}J^\alpha u = v = J^{\alpha-\beta}u$. Thus (2.24) is proved. Thus the proof of Theorem 2.3 is completed. ∎

Now we define the fractional derivative ∂_t^α in $H_\alpha(0, T)$.

Definition 2.1 For $0 \le \alpha \le 1$, we set

$$\partial_t^\alpha u := J^{-\alpha} u, \quad u \in H_\alpha(0, T)$$

with the domain $\mathcal{D}(\partial_t^\alpha) = H_\alpha(0, T)$.

We recall that, $\mathcal{D}(\partial_t^\alpha)$ denotes the domain of the operator ∂_t^α.

Remark 2.4 In this book, we mainly consider ∂_t^α for the case of $0 < \alpha < 1$. For $\alpha > 1$, $\notin \mathbb{N}$, on the basis of ∂_t^α with $0 < {}^{\cdot}\alpha < 1$, we can define as follows: Let $\alpha = m + \gamma$ with $m \in \mathbb{N}$ and $0 < \gamma < 1$. Then

$$\partial_t^\alpha u = \partial_t^\gamma \left(\frac{d^m u}{dt^m} \right)$$

with

$$\mathcal{D}(\partial_t^\alpha) = \left\{ u \in H^m(0, T); \ u(0) = \cdots = \frac{d^{m-1}u}{dt^{m-1}}(0) = 0, \ \frac{d^m u}{dt^m} \in H_\gamma(0, T) \right\}$$

(cf. (3.39)), and we can argue the isomorphism and fractional differential equations in the same way as the later parts of this book, but we omit the details and postpone them to forthcoming publications.

By Theorem 2.1, we note that $H_\alpha(0, T) = J^\alpha L^2(0, T)$. Therefore ∂_t^α in $H_\alpha(0, T)$ is well-defined and $\partial_t^\alpha u \in L^2(0, T)$ for $u \in H_\alpha(0, T)$. Moreover, $\partial_t^\alpha : H_\alpha(0, T) \longrightarrow L^2(0, T)$ is surjective. Indeed, let $v \in L^2(0, T)$ be arbitrarily given. By Theorem 2.1, we have $\varphi := J^\alpha v \in H_\alpha(0, T)$ and so $\partial_t^\alpha \varphi = v$ by the definition, which means that $\partial_t^\alpha : H_\alpha(0, T) \longrightarrow L^2(0, T)$ is surjective.

On the other hand, we can prove

$$J^{-1} u = \frac{du}{dt}, \quad u \in H_1(0, T).$$

Indeed, setting $v = J^{-1}u$, we have $v \in L^2(0, T)$ by Theorem 2.1 and $u = Jv$, that is, $u(t) = \int_0^t v(s)ds$. By $v \in L^2(0, T)$, we see that $u \in AC[0, T]$, that is, u is absolutely continuous on $[0, T]$ and $\frac{du}{dt}(t) = v(t)$ for almost all $t \in (0, T)$, that is, $\frac{du}{dt}(t) = (J^{-1}u)(t)$ for almost all $t \in (0, T)$. Replacing α and β respectively by $1 - \alpha$ and 1 in Theorem 2.3 (ii), we obtain

$$J^{-\alpha} = J^{(1-\alpha)-1} = J^{-1}J^{1-\alpha},$$

that is, $J^{-\alpha} = \frac{d}{dt}(J^{1-\alpha})$.

Noting that $J^{-\alpha} = J^{-1}J^{1-\alpha}$ and summing up, we can state:

Theorem 2.4 *Let $0 < \alpha < 1$. Then ∂_t^α is an isomorphism between $H_\alpha(0, T)$ and $L^2(0, T)$. That is, $\partial_t^\alpha : H_\alpha(0, T) \longrightarrow L^2(0, T)$ is injective and surjective, and*

$$\|\partial_t^\alpha u\|_{L^2(0,T)} \sim \|u\|_{H_\alpha(0,T)}. \tag{2.25}$$

Moreover,

$$\partial_t^\alpha u = J^{-\alpha}u = \frac{d}{dt}(J^{1-\alpha}u) = D_t^\alpha u, \quad u \in H_\alpha(0, T) \tag{2.26}$$

and

$$\partial_t^\alpha u = D_t^\alpha u = d_t^\alpha u \quad for\ u \in H_\alpha(0, T) \cap {}_0W^{1,1}(0, T). \tag{2.27}$$

Indeed, (2.27) follows from (2.26) and Lemma 1.3. Thus we can calculate $\partial_t^\alpha u$ by means of $D_t^\alpha u$ for $u \in H_\alpha(0, T)$.

We are tempted to assert (2.27) for $u \in H_\alpha(0, T)$. However, $d_t^\alpha u$ does not directly make sense for $u \in H_\alpha(0, T)$ and with suitable extension of d_t^α, we can transfer (2.27) to $H_\alpha(0, T)$ (see Theorem 2.5).

Formula $w = \partial_t^\alpha u = \frac{d}{dt}(J^{1-\alpha}u)$ in (2.26) can correspond to the classical inversion for finding w solving $J^\alpha w = u$ (e.g., Gorenflo and Vessella [8]) for $u \in {}_0W^{1,1}(0, T)$, but our construction for ∂_t^α guarantees the formula for $u \in H_\alpha(0, T)$, which is a wider space than the set of all absolutely continuous functions on $[0, T]$. Moreover,

Proposition 2.1

$$J^\alpha D_t^\alpha u(t) = J^\alpha d_t^\alpha u(t) = u(t)$$

for $u \in {}_0W^{1,1}(0, T)$.

This proposition means that J^α is the left inverse to D_t^α and d_t^α for $u \in {}_0W^{1,1}(0, T)$.

Proof By Lemma 1.3, we have

$$J^{1-\alpha}\frac{du}{dt}(t) = d_t^\alpha u(t) = D_t^\alpha u(t), \quad u \in {}_0W^{1,1}(0, T).$$

Therefore

$$J^\alpha J^{1-\alpha}\frac{du}{dt}(t) = J^\alpha d_t^\alpha u(t) = J^\alpha D_t^\alpha u(t), \quad u \in {}_0W^{1,1}(0, T).$$

By $J^\alpha J^{1-\alpha} = J$ we have

$$J^\alpha J^{1-\alpha} \frac{du}{dt}(t) = J \frac{du}{dt}(t) = \int_0^t \frac{du}{ds}(s)ds = u(t)$$

as $u(0) = 0$. Thus the proof of the proposition is completed. ∎

Henceforth we regard ∂_t^α as an operator with the domain $\mathcal{D}(\partial_t^\alpha) = H_\alpha(0, T)$ if we do not specify. We can estimate $\|\partial_t^\alpha u\|_{L^2(0,T)}$ by the Sobolev norm but our theorem establishes also the surjectivity for the convenience for applications to fractional differential equations.

We conclude this section with the equivalence of ∂_t^α with the closed extension of the Caputo derivative operator d_t^α. As for the closed extension and the closure of an operator, see e.g., Kato [13] (Chapter III, §5). We consider the classical Caputo derivative

$$d_t^\alpha u(t) = \frac{1}{\Gamma(1 - \alpha)} \int_0^t (t - s)^{-\alpha} \frac{du}{ds}(s)ds$$

with $\mathcal{D}(d_t^\alpha) = {}_0C^1[0, T]$. We consider d_t^α as an operator from $\mathcal{D}(d_t^\alpha) = {}_0C^1[0, T] \subset L^2(0, T)$ to $L^2(0, T)$.

By $\overline{d_t^\alpha}$ we denote the closure in $L^2(0, T)$ of d_t^α with $\mathcal{D}(d_t^\alpha) = {}_0C^1[0, T]$, which is the smallest closed extension of d_t^α. Then we prove:

Theorem 2.5 *We have* $\mathcal{D}(\overline{d_t^\alpha}) = H_\alpha(0, T)$, *and*

$$\overline{d_t^\alpha} = \partial_t^\alpha = D_t^\alpha \quad on \ H_\alpha(0, T).$$

This theorem means that our definition of ∂_t^α is consistent with the classical Caputo derivative by considering the closure of the operator.

Proof of Theorem 2.5 We first prove that the operator d_t^α is closable.
Indeed, let $u_n \in {}_0C^1[0, T]$ and $u_n \longrightarrow 0$ in $L^2(0, T)$ and $d_t^\alpha u_n \longrightarrow v$ in $L^2(0, T)$ with some $v \in L^2(0, T)$. Then by Lemma 1.3 and Theorem 2.4, we have $d_t^\alpha u_n = \partial_t^\alpha u_n$, $n = 1, 2, 3, \ldots$ Therefore $\partial_t^\alpha u_n \longrightarrow v$ in $L^2(0, T)$. Theorem 2.4 yields that $u_n \in {}_0C^1[0, T] \subset H_\alpha(0, T)$ is a Cauchy sequence in $H_\alpha(0, T)$. Therefore there exists some $u \in H_\alpha(0, T)$ such that $u_n \longrightarrow u$ in $H_\alpha(0, T)$. Since $u_n \longrightarrow 0$ in $L^2(0, T)$, we see that $u = 0$, that is, $u_n \longrightarrow 0$ in $H_\alpha(0, T)$. Hence again by Theorem 2.4, we obtain $\partial_t^\alpha u_n = d_t^\alpha u_n \longrightarrow \partial_t^\alpha 0 = 0$, that is, $v = 0$. Thus d_t^α is closable.

Now we return to the proof of Theorem 2.5. We recall that $u \in \mathcal{D}(\overline{d_t^\alpha})$ if and only if there exist $u_n \in {}_0C^1[0, T]$ such that $u_n \longrightarrow u$ in $L^2(0, T)$ and $d_t^\alpha u_n$ is a Cauchy sequence in $L^2(0, T)$. Since $d_t^\alpha u_n = \partial_t^\alpha u_n$ for $u_n \in {}_0C^1[0, T]$ by Lemma 1.3 and Theorem 2.4, we see that u_n is a Cauchy sequence in $H_\alpha(0, T)$. Therefore there exists $\widetilde{u} \in H_\alpha(0, T)$ such that $u_n \longrightarrow \widetilde{u}$ in $H_\alpha(0, T)$ by Theorem 2.4. Since $u_n \longrightarrow u$ in $L^2(0, T)$, we obtain $\widetilde{u} = u$ and $u_n \longrightarrow u$ in $H_\alpha(0, T)$ and $d_t^\alpha u_n \longrightarrow \partial_t^\alpha u$ in

$L^2(0, T)$. The definition of $\overline{d_t^\alpha}$ yields

$$\overline{d_t^\alpha} u = \lim_{n \to \infty} d_t^\alpha u_n = \lim_{n \to \infty} \partial_t^\alpha u_n = \partial_t^\alpha u.$$

Hence we proved that $\mathcal{D}(\overline{d_t^\alpha}) \subset H_\alpha(0, T)$ and $\overline{d_t^\alpha} u = \partial_t^\alpha u$ for $u \in \mathcal{D}(\overline{d_t^\alpha})$.

Conversely assume that $u \in H_\alpha(0, T)$. Since $\overline{{}_0C^1[0, T]}^{H_\alpha(0,T)} = H_\alpha(0, T)$ by Lemma 2.2, there exist $u_n \in {}_0C^1[0, T]$ such that $u_n \longrightarrow u$ in $H_\alpha(0, T)$. Then $\partial_t^\alpha u_n = d_t^\alpha u_n \longrightarrow \partial_t^\alpha u$ in $L^2(0, T)$ by Theorem 2.4. That is, $u_n \in {}_0C^1[0, T]$ is d_t^α-convergent to u and so $u \in \mathcal{D}(\overline{d_t^\alpha})$ and $\overline{d_t^\alpha} u = \partial_t^\alpha u$. Thus the proof of Theorem 2.5 is complete. ∎

Remark 2.5 In Sect. 2.5, we discuss the case where we start the operator d_t^α with $\mathcal{D}(d_t^\alpha) = C^1[0, T]$, not ${}_0C^1[0, T]$.

2.4 Some Functions in $H_\alpha(0, T)$

In this section, we show that the function t^γ and some functions defined by the Mittag-Leffler functions are in $H_\alpha(0, T)$. Some of these inclusions are used in later sections.

For $\alpha, \beta > 0$, we define

$$\begin{cases} E_{\alpha,\beta}(z) = \sum_{k=0}^\infty \frac{z^k}{\Gamma(\alpha k+\beta)}, \\ E_{\alpha,1}(z) = \sum_{k=0}^\infty \frac{z^k}{\Gamma(\alpha k+1)}, \quad z \in \mathbb{C}. \end{cases} \tag{2.28}$$

These functions are called the Mittag-Leffler functions and play important roles in the fractional calculus (e.g., [15, 24]).

It is known that $E_{\alpha,\beta}(z)$ is an entire function in z with $\alpha, \beta > 0$. The Mittag-Leffler functions have been well studied and here we describe only a few of the important properties used later.

Lemma 2.5

(i) *Let $0 < \alpha < 2$ and $\beta > 0$. We assume that $\frac{\pi\alpha}{2} < \mu < \min\{\pi, \pi\alpha\}$. Then there exists a constant $C = C(\alpha, \beta, \mu) > 0$ such that*

$$|E_{\alpha,\beta}(z)| \le \frac{C}{1 + |z|}, \quad \mu \le |\arg z| \le \pi.$$

(ii) *For $\lambda \in \mathbb{R}$ and $\alpha > 0$, $m \in \mathbb{N}$, we have*

$$\frac{d^m}{dt^m} E_{\alpha,1}(-\lambda t^\alpha) = -\lambda t^{\alpha-m} E_{\alpha,\alpha-m+1}(-\lambda t^\alpha), \quad t > 0.$$

The proof of (i) can be found e.g., in [24]. The proof of (ii) is seen directly because $E_{\alpha,1}(z)$ is entire in z and we can differentiate $E_{\alpha,1}(-\lambda t^\alpha) = \sum_{k=0}^\infty \frac{(-\lambda t^\alpha)^k}{\Gamma(\alpha k+1)}$ termwise.

Proposition 2.2 *Let $0 < \alpha < 1$ and $\lambda \in \mathbb{R}$. Then*

$$E_{\alpha,1}(-\lambda t^\alpha) - 1 \in W_\alpha(0, T) \subset {}_0W^{1,1}(0, T) \cap H_\alpha(0, T).$$

Here we recall that $W_\alpha(0, T)$ is defined by (2.5).

Proof The inclusion $W_\alpha(0, T) \subset H_\alpha(0, T)$ is proved already in Lemma 2.1. Moreover, Lemma 2.5 yields

$$\frac{d}{dt} E_{\alpha,1}(-\lambda t^\alpha) = -\lambda t^{\alpha-1} E_{\alpha,\alpha}(-\lambda t^\alpha)$$

and

$$\left| \frac{d}{dt} E_{\alpha,1}(-\lambda t^\alpha) \right| \le |\lambda| t^{\alpha-1} |E_{\alpha,\alpha}(-\lambda t^\alpha)| \le C |\lambda| t^{\alpha-1}, \quad t > 0.$$

By $E_{\alpha,1}(0) - 1 = \frac{1}{\Gamma(1)} - 1 = 0$, the definition of $W_\alpha(0, T)$ implies the conclusion. ∎

Proposition 2.3 *Let $0 < \alpha < 1$. Then*

$$\int_0^t (t-s)^{\alpha-1} E_{\alpha,\alpha}(-\lambda(t-s)^\alpha) f(s) ds \in H_\alpha(0, T)$$

and

$$\left\| \int_0^t (t-s)^{\alpha-1} E_{\alpha,\alpha}(-\lambda(t-s)^\alpha) f(s) ds \right\|_{H_\alpha(0,T)} \le C \|f\|_{L^2(0,T)}$$

for all $f \in L^2(0, T)$.

For $f \in L^2(0, T)$ we cannot expect more regularity than $H_\alpha(0, T)$. For $f \in L^2(0, T)$, we set

$$v(t) := \int_0^t (t-s)^{\alpha-1} E_{\alpha,\alpha}(-\lambda(t-s)^\alpha) f(s) ds,$$

Then $\partial_t^\alpha v$ is well-defined by $v \in H_\alpha(0, T)$, and in Sect. 3.4 of Chap. 3, we prove that v satisfies

$$\partial_t^\alpha v(t) = -\lambda v(t) + f(t), \quad 0 < t < T$$

(see Proposition 3.2).

We can generalize the proposition: if $f \in L^2(0, T)$ and G is an entire function, then

$$\int_0^t (t-s)^{\alpha-1} G((t-s)^\alpha) f(s) ds \in H_\alpha(0, T).$$

Proof We set

$$g_k(s) := \frac{(-\lambda)^k s^{\alpha k + \alpha - 1}}{\Gamma(\alpha(k+1))}, \quad k = 0, 1, 2, 3, .., \quad R(s) = \sum_{k=N}^{\infty} g_k(s).$$

Then

$$(t-s)^{\alpha-1} E_{\alpha,\alpha}(-\lambda(t-s)^\alpha) = \sum_{k=0}^{\infty} \frac{(-\lambda)^k (t-s)^{\alpha k + \alpha - 1}}{\Gamma(\alpha(k+1))} = \sum_{k=0}^{N-1} g_k(t-s) + R(t-s).$$

$$(2.29)$$

We choose $N \in \mathbb{N}$ such that

$$\alpha N > \frac{3}{2} - \alpha. \tag{2.30}$$

Then

$$\alpha k + \alpha - 1 \geq \alpha N + \alpha - 1 > 0 \quad \text{if } k \geq N,$$

and so $R \in C[0, \infty)$ and $R(0) = 0$. By the termwise differentiation, we obtain

$$\frac{d}{ds} R(s) = \sum_{k=N}^{\infty} \frac{(-\lambda)^k (\alpha k + \alpha - 1) s^{\alpha k + \alpha - 2}}{\Gamma(\alpha(k+1))} = s^{\alpha N + \alpha - 2} \sum_{k=N}^{\infty} \frac{(-\lambda)^k s^{\alpha(k-N)}}{\Gamma(\alpha(k+1) - 1)}$$

$$= s^{\alpha N + \alpha - 2} \sum_{j=0}^{\infty} \frac{(-\lambda)^{N+j} s^{\alpha j}}{\Gamma(\alpha N + \alpha - 1 + \alpha j)} = (-\lambda)^N s^{\alpha N + \alpha - 2} E_{\alpha, \alpha N + \alpha - 1}(-\lambda s^\alpha).$$

Since $\alpha N + \alpha - 2 > -1$ by (2.30), we see that $\frac{dR}{ds} \in L^1(0, T)$. Therefore

$$\frac{d}{dt} \int_0^t R(t-s) f(s) ds = \int_0^t \frac{dR}{dt} (t-s) f(s) ds,$$

so that the Young inequality for the convolution (Lemma A.1 in the Appendix) yields

$$\left\| \frac{d}{dt} \int_0^t R(t-s) f(s) ds \right\|_{L^2(0,T)} \leq C \|f\|_{L^2(0,T)},$$

which means that $\int_0^t R(t-s)f(s)ds \in W^{1,1}(0, T)$. Moreover,

$$\left|\frac{d}{dt}\int_0^t R(t-s)f(s)ds\right| \leq C\int_0^t (t-s)^{\alpha N+\alpha-2}|f(s)|ds.$$

Since $\alpha N > \frac{3}{2} - \alpha$ by (2.30), we have $2\alpha N + 2\alpha - 4 > -1$. Therefore the Cauchy–Schwarz inequality and (2.30) yield

$$\left|\frac{d}{dt}\int_0^t R(t-s)f(s)ds\right| \leq C\left(\int_0^t (t-s)^{2\alpha N+2\alpha-4}ds\right)^{\frac{1}{2}}\|f\|_{L^2(0,T)}$$

$$= \frac{C}{(2\alpha N + 2\alpha - 3)^{\frac{1}{2}}}t^{\alpha N+\alpha-\frac{3}{2}}\|f\|_{L^2(0,T)} \leq C_1\|f\|_{L^2(0,T)}.$$

Again by (2.30), we see that $\alpha N + \alpha - 1 > 0$, so that $\int_0^t R(t-s)f(s)ds$ vanishes at $t = 0$. Hence we proved

$$\int_0^t R(t-s)f(s)ds \in {}_0W^{1,1}(0, T) \cap W_\alpha(0, T) \subset {}_0W^{1,1}(0, T) \cap H_\alpha(0, T).$$

Therefore by (2.27) we have

$$\partial_t^\alpha\left(\int_0^t R(t-s)f(s)ds\right) = \frac{1}{\Gamma(1-\alpha)}\int_0^t (t-s)^{-\alpha}\left(\frac{d}{ds}\int_0^s R(s-\xi)f(\xi)d\xi\right)ds.$$

Hence

$$\left|\partial_t^\alpha\left(\int_0^t R(t-s)f(s)ds\right)\right| \leq C\int_0^t (t-s)^{-\alpha}\left|\frac{d}{ds}\int_0^s R(s-\xi)f(\xi)d\xi\right|ds$$

$$\leq C\int_0^t (t-s)^{-\alpha}ds \times C_1\|f\|_{L^2(0,T)} \leq CT^{1-\alpha}\|f\|_{L^2(0,T)}.$$

Therefore

$$\left\|\partial_t^\alpha\left(\int_0^t R(t-s)f(s)ds\right)\right\|_{L^2(0,T)}^2 \leq \int_0^T C^2 T^{2-2\alpha}\|f\|_{L^2(0,T)}^2 dt,$$

that is, (2.25) yields

$$\left\|\int_0^t R(t-s)f(s)ds\right\|_{H_\alpha(0,T)} \leq C\|f\|_{L^2(0,T)}. \tag{2.31}$$

Moreover, by the definition (1.5) of J^α, the Eq. (2.29) implies

$$\int_0^t (t-s)^{\alpha-1} E_{\alpha,\alpha}(-\lambda(t-s)^\alpha) f(s) ds$$

$$= \sum_{k=0}^{N-1} (-\lambda)^k (J_{\alpha k+\alpha} f)(t) + \int_0^t R(t-s) f(s) ds.$$

Since $0 < \alpha k + \alpha \leq \alpha N$ for $0 \leq k \leq N-1$, Theorem 2.1 yields

$$J_{\alpha k+\alpha} f \in \begin{cases} H_{\alpha k+\alpha}(0,T), & \alpha k + \alpha < 1, \\ H_1(0,T), & \alpha k + \alpha \geq 1, \; k = 0, \ldots, N-1 \end{cases}$$

and so

$$\sum_{k=0}^{N-1} (-\lambda)^k J_{\alpha k+\alpha} f \in \bigcap_{k=0}^{N-1} H_{\alpha k+\alpha}(0,T) \cap H_1(0,T) = H_\alpha(0,T).$$

Thus with (2.31), the proof of Proposition 2.3 is complete.   ∎

Moreover, we can prove:

Proposition 2.4 Let $0 < \alpha < \frac{1}{2}$ and $\delta > 0$ satisfy $\alpha + \delta - \frac{1}{2} < 0$. Then

$$t^{\alpha+\delta-\frac{1}{2}} \in H_\alpha(0,T) = H^\alpha(0,T).$$

Proof In view of Theorem 2.1, it suffices to verify

$$t^{\alpha+\delta-\frac{1}{2}} = J^\alpha w_0(t),$$

where

$$w_0(t) := \frac{\Gamma\left(\alpha + \delta + \frac{1}{2}\right)}{\Gamma\left(\delta + \frac{1}{2}\right)} t^{\delta-\frac{1}{2}} \in L^2(0,T),$$

but we prove directly. We set

$$\theta = -\alpha - \delta + \frac{1}{2}.$$

Then $0 < \theta < \frac{1}{2} - \alpha$. It suffices to prove

$$\int_0^T \int_0^T \frac{|t^{-\theta} - s^{-\theta}|^2}{|t-s|^{1+2\alpha}} ds dt < \infty. \tag{2.32}$$

Using

$$\int_0^T \left(\int_t^T \cdots ds \right) dt = \int_0^T \left(\int_0^s \cdots dt \right) ds,$$

similarly to (2.7), we have

$$\int_0^T \int_0^T \frac{|t^{-\theta} - s^{-\theta}|^2}{|t-s|^{1+2\alpha}} ds dt = \int_0^T \left(\int_0^t \frac{|t^{-\theta} - s^{-\theta}|^2}{|t-s|^{1+2\alpha}} ds \right) dt + \int_0^T \left(\int_t^T \frac{|t^{-\theta} - s^{-\theta}|^2}{|t-s|^{1+2\alpha}} ds \right) dt$$

$$= 2 \int_0^T \left(\int_0^t \frac{|t^{-\theta} - s^{-\theta}|^2}{|t-s|^{1+2\alpha}} ds \right) dt.$$

We can prove

$$\int_0^T \left(\int_0^t s^{\gamma_1}(t-s)^{\gamma_2} t^{\gamma_3} ds \right) dt < \infty \tag{2.33}$$

if $\gamma_1, \gamma_2 > -1$ and $\gamma_1 + \gamma_2 + \gamma_3 > -2$. ∎

Proof of (2.33) By $\gamma_1, \gamma_2 > -1$, we have

$$\int_0^T \left(\int_0^t s^{\gamma_1}(t-s)^{\gamma_2} ds \right) t^{\gamma_3} dt = \frac{\Gamma(1+\gamma_1)\Gamma(1+\gamma_2)}{\Gamma(2+\gamma_1+\gamma_2)} \int_0^T t^{\gamma_1+\gamma_2+1} t^{\gamma_3} dt,$$

which is convergent by $\gamma_1 + \gamma_2 + \gamma_3 + 1 > -1$. Thus the proof of (2.33) is complete. ∎

Now we will prove

$$\int_0^T \int_0^t \frac{|t^{-\theta} - s^{-\theta}|^2}{|t-s|^{1+2\alpha}} ds dt = \int_0^T \int_0^t t^{-2\theta} s^{-2\theta} \frac{|t^\theta - s^\theta|^2}{|t-s|^{1+2\alpha}} ds dt < \infty. \tag{2.34}$$

We choose $0 < \mu < 1$ such that

$$\frac{1}{2-2\theta} < \mu < \min\left\{ 1, \frac{1-\alpha}{1-\theta} \right\}. \tag{2.35}$$

This is possible because $0 < \frac{1}{2-2\theta} < \min\left\{ 1, \frac{1-\alpha}{1-\theta} \right\}$ by $0 < \theta < \frac{1}{2}$ and $0 < \alpha < \frac{1}{2}$. Indeed, $\frac{1}{2-2\theta} < 1$ is clearly seen, and $\frac{1-\alpha}{1-\theta} - \frac{1}{2-2\theta} = \frac{1-2\alpha}{2-2\theta} > 0$ by $0 < \alpha < \frac{1}{2}$.

By $0 < \theta < 1$, we can easily see that $a^\theta + b^\theta \geq (a+b)^\theta$ for $a, b \geq 0$ and so $|t^\theta - s^\theta| \leq |t-s|^\theta$. Hence

$$|t^\theta - s^\theta|^{2\mu} \leq |t-s|^{2\mu\theta}$$

for $0 \leq s \leq t$. On the other hand, the mean value theorem and $\theta < 1$ yield

$$|t^\theta - s^\theta| \leq \max_{s \leq \eta \leq t} \theta \eta^{\theta-1}(t-s) \leq \theta s^{\theta-1}(t-s)$$

for $0 \leq s \leq t$. Therefore

$$|t^\theta - s^\theta|^{2(1-\mu)} \leq C s^{2(1-\mu)(\theta-1)} |t - s|^{2(1-\mu)},$$

and so

$$|t^\theta - s^\theta|^2 = |t^\theta - s^\theta|^{2\mu} |t^\theta - s^\theta|^{2-2\mu} \leq C s^{2(1-\mu)(\theta-1)} (t - s)^{2(1-\mu)+2\mu\theta}.$$

Hence

$$\int_0^T \int_0^t \frac{|t^\theta - s^\theta|^2 t^{-2\theta} s^{-2\theta}}{|t - s|^{1+2\alpha}} ds dt \leq C \int_0^T \int_0^t s^{2(1-\mu)(\theta-1)-2\theta} t^{-2\theta} |t - s|^{1-2\alpha-2\mu+2\mu\theta} ds dt.$$

By $0 < \alpha < \frac{1}{2}, 0 < \theta < \frac{1}{2} - \alpha$ and (2.35), we directly verify

$$2(1 - \mu)(\theta - 1) - 2\theta > -1, \quad 1 - 2\alpha - 2\mu + 2\mu\theta > -1$$

and

$$\{2(1 - \mu)(\theta - 1) - 2\theta\} + (-2\theta) + (1 - 2\alpha - 2\mu + 2\mu\theta) > -2,$$

so that (2.33) yields (2.34). Thus the proof of Proposition 2.4 is complete. ∎

Remark 2.6 It seems that we cannot directly prove the lemma without introducing the parameter $\mu \in (0, 1)$.

Since Proposition 2.4 proves $C_0 t^{\alpha+\delta-\frac{1}{2}} \in H_\alpha(0, T)$, where $C_0 = \dfrac{\Gamma\left(\delta+\frac{1}{2}\right)}{\Gamma\left(\alpha+\delta+\frac{1}{2}\right)}$, we can apply Theorem 2.5 to calculate

$$\partial_t^\alpha (C_0 t^{\alpha+\delta-\frac{1}{2}}) = \overline{d_t^\alpha}(C_0 t^{\alpha+\delta-\frac{1}{2}}) = D_t^\alpha (C_0 t^{\alpha+\delta-\frac{1}{2}})$$

$$= \frac{C_0}{\Gamma(1-\alpha)} \frac{d}{dt} \int_0^t (t-s)^{-\alpha} s^{\alpha+\delta-\frac{1}{2}} ds$$

$$= \frac{C_0}{\Gamma(1-\alpha)} \frac{\Gamma(1-\alpha)\Gamma\left(\alpha+\delta+\frac{1}{2}\right)}{\Gamma\left(\delta+\frac{3}{2}\right)} \frac{d}{dt} t^{\delta+\frac{1}{2}} = t^{\delta-\frac{1}{2}}.$$

Here we used $\Gamma\left(\delta+\frac{3}{2}\right) = \left(\delta+\frac{1}{2}\right)\Gamma\left(\delta+\frac{1}{2}\right)$.

Therefore with the closure of d_t^α defined in Sect. 2.3 and ∂_t^α, we can justify

$$\overline{d_t^\alpha}(C_0 t^{\alpha+\delta-\frac{1}{2}}) = t^{\delta-\frac{1}{2}} \tag{2.36}$$

for $0 < \alpha < \frac{1}{2}$ and small $\delta > 0$ satisfying $\alpha + \delta - \frac{1}{2} < 0$.

2.5 Definition of ∂_t^α in $H^\alpha(0, T)$

We have defined $\partial_t^\alpha u$ for $u \in H_\alpha(0, T)$. Next, it is natural to define the fractional derivative in $H^\alpha(0, T)$ which is wider than $H_\alpha(0, T)$ for $0 < \alpha < 1$, although we work mainly within $H_\alpha(0, T)$ for discussing fractional differential equations.

Since $H_\alpha(0, T)$ is a set of functions in $H^\alpha(0, T)$ which "vanish" in some sense at $t = 0$, the following definition is naive but may be reasonable for $0 < \alpha < 1$ with $\alpha \neq \frac{1}{2}$: we define $\partial_t^\alpha v$ for $v \in H^\alpha(0, T)$ by

$$\partial_t^\alpha v = \begin{cases} \partial_t^\alpha v, & v \in H^\alpha(0, T), \ 0 < \alpha < \frac{1}{2}, \\ \partial_t^\alpha(v - v(0)), & v \in H^\alpha(0, T), \ \frac{1}{2} < \alpha < 1. \end{cases} \tag{2.37}$$

Then $\partial_t^\alpha u$ is well-defined because $H_\alpha(0, T) = H^\alpha(0, T)$ for $0 < \alpha < \frac{1}{2}$, and $H^\alpha(0, T) \subset C[0, T]$ for $\frac{1}{2} < \alpha < 1$ which enables us to define $v(0)$.

For $0 < \alpha < 1$ and $\alpha \neq \frac{1}{2}$, we note that this definition gives the same result for $v \in H_\alpha(0, T)$, and so we can extend the domain of ∂_t^α from $H_\alpha(0, T)$ to $H^\alpha(0, T)$.

On the other hand, the case of $\alpha = \frac{1}{2}$ is delicate, and we do not define $\partial_t^{\frac{1}{2}} u$ for $u \in H^{\frac{1}{2}}(0, T) \supsetneq H_{\frac{1}{2}}(0, T)$. For example, we note that $1 \in H^{\frac{1}{2}}(0, T)$ but $1 \notin H_{\frac{1}{2}}(0, T)$. We do not define $\partial_t^{\frac{1}{2}} 1$ in $L^2(0, T)$, although we can calculate $d_t^{\frac{1}{2}} 1 = 0$ and $D_t^{\frac{1}{2}} 1 = \frac{1}{\Gamma(\frac{1}{2})} t^{-\frac{1}{2}}$ pointwise.

By the definition we note that

$$\partial_t^\alpha(u_1 + u_2) = \partial_t^\alpha u_1 + \partial_t^\alpha u_2, \quad u_1, u_2 \in H^\alpha(0, T)$$

for $0 < \alpha < 1$ and $\alpha \neq \frac{1}{2}$.

Throughout this book, we always interpret $\partial_t^\alpha u$ with $0 < \alpha < 1$ and $\alpha \neq \frac{1}{2}$ in the sense of (2.37) for $u \in H^\alpha(0, T)$.

Now we show properties of ∂_t^α in conventional Sobolev space $H^\alpha(0, T)$.

Proposition 2.5

(i)

$$\partial_t^\alpha 1 = \begin{cases} \frac{t^{-\alpha}}{\Gamma(1-\alpha)}, & 0 < \alpha < \frac{1}{2}, \\ 0, & \frac{1}{2} < \alpha < 1. \end{cases} \tag{2.38}$$

(ii)

$$\partial_t^\alpha u = \begin{cases} D_t^\alpha u, & 0 < \alpha < \frac{1}{2}, \\ d_t^\alpha u, & \frac{1}{2} < \alpha < 1 \end{cases} \tag{2.39}$$

for $u \in C^1[0, T]$.

(iii) Let $\frac{1}{2} < \alpha < 1$ and $u \in H^\alpha(0, T)$. Then there exists a sequence $\varphi_k \in C^1[0, T]$,
    $k \in \mathbb{N}$ such that $\varphi_k \longrightarrow u$ in $H^\alpha(0, T)$ and $\partial_t^\alpha u = \lim_{k \to \infty} \partial_t^\alpha \varphi_k$ in $L^2(0, T)$.
(iv) Let $0 < \alpha < \frac{1}{2}$. Then there exists a constant $C > 0$ such that

$$C^{-1} \|u - a\|_{H^\alpha(0,T)} \leq \|\partial_t^\alpha u\|_{L^2(0,T)} + |a| \leq C(\|u - a\|_{H^\alpha(0,T)} + |a|)$$

    for all $a \in \mathbb{R}$ and $u \in H^\alpha(0, T)$.
(v) Let $\frac{1}{2} < \alpha < 1$. Then there exists a constant $C > 0$ such that

$$C^{-1} \|u\|_{H^\alpha(0,T)} \leq \|\partial_t^\alpha u\|_{L^2(0,T)} + |u(0)| \leq C\|u\|_{H^\alpha(0,T)}$$

    for all $u \in H^\alpha(0, T)$.

Proof

(i) By (2.37), setting $u(t) \equiv 1$, we have $\partial_t^\alpha 1 = \partial_t^\alpha(u - u(0)) = \partial_t^\alpha(1 - 1) = \partial_t^\alpha 0 = 0$ for $\frac{1}{2} < \alpha < 1$. For $0 < \alpha < \frac{1}{2}$, by (2.26) we can directly calculate

$$\partial_t^\alpha 1 = \frac{d}{dt}(J^{1-\alpha} 1) = \frac{d}{dt} \frac{1}{\Gamma(1-\alpha)} \int_0^t (t-s)^{-\alpha} ds$$

$$= \frac{d}{dt}\left(\frac{1}{(1-\alpha)\Gamma(1-\alpha)} t^{1-\alpha}\right) = \frac{t^{-\alpha}}{\Gamma(1-\alpha)}.$$

This completes the proof of (i).

(ii) Since $u \in C^1[0, T] \subset H^\alpha(0, T) = H_\alpha(0, T)$ for $0 < \alpha < \frac{1}{2}$ and u is absolutely continuous, equality (2.26) yields that $\partial_t^\alpha u = D_t^\alpha u$ for $0 < \alpha < \frac{1}{2}$. Next we verify (2.39) for $\frac{1}{2} < \alpha < 1$. By (2.37), we have $\partial_t^\alpha u = \partial_t^\alpha(u - u(0))$. Applying Lemma 1.1 to $u - u(0)$ and noting that $u \in C^1[0, T]$, we obtain

$$\partial_t^\alpha(u - u(0)) = d_t^\alpha(u - u(0)) = \frac{1}{\Gamma(1-\alpha)} \int_0^t (t-s)^{-\alpha} \frac{d}{ds}(u(s) - u(0)) ds$$

$$= \frac{1}{\Gamma(1-\alpha)} \int_0^t (t-s)^{-\alpha} \frac{du}{ds}(s) ds = d_t^\alpha u(t).$$

(iii) Since $C^1[0, T]$ is dense in $H^\alpha(0, T)$, there exists a sequence $\varphi_k \in C^1[0, T]$, $k \in \mathbb{N}$ such that $\varphi_k \longrightarrow u$ in $H^\alpha(0, T)$. By $\alpha > \frac{1}{2}$, the Sobolev embedding yields $\varphi_k(0) \longrightarrow u(0)$. Therefore $\varphi_k - \varphi_k(0) \in {}_0C^1[0, T]$ and $\varphi_k - \varphi_k(0) \longrightarrow u - u(0)$ in $H_\alpha(0, T)$. By the definition (2.37), we see that

$$\partial_t^\alpha u = \partial_t^\alpha(u - u(0)) = \lim_{k \to \infty} \partial_t^\alpha(\varphi_k - \varphi_k(0)) = \lim_{k \to \infty} \partial_t^\alpha \varphi_k$$

in $L^2(0, T)$.

(iv) For $0 < \alpha < \frac{1}{2}$, we note that $H_\alpha(0,T) = H^\alpha(0,T)$, and so $u \in H^\alpha(0,T)$ implies $u - a \in H_\alpha(0,T)$. Therefore, by Theorem 2.2, we have

$$\|u - a\|_{H^\alpha(0,T)} \sim \|J^{-\alpha}(u-a)\|_{L^2(0,T)} = \|\partial_t^\alpha u - \partial_t^\alpha a\|_{L^2(0,T)} \qquad (2.40)$$

if $u - a \in H_\alpha(0,T)$. Hence (2.38) and (2.40) imply

$$\|u - a\|_{H^\alpha(0,T)} \geq \|u\|_{H^\alpha(0,T)} - \|a\|_{H^\alpha(0,T)}$$

$$\geq C_1 \|\partial_t^\alpha u\|_{L^2(0,T)} - C_2 \|\partial_t^\alpha a\|_{L^2(0,T)}$$

$$= C_1 \|\partial_t^\alpha u\|_{L^2(0,T)} - \frac{C_2}{\Gamma(1-\alpha)} \|t^{-\alpha}\|_{L^2(0,T)} |a| \geq C_1 \|\partial_t^\alpha u\|_{L^2(0,T)} - C_3 |a|,$$

which is the second inequality in (iv). Next, (2.40) yields

$$\|u - a\|_{H^\alpha(0,T)} \leq \|u\|_{H^\alpha(0,T)} + \|a\|_{H^\alpha(0,T)}$$

$$\leq C_4(\|\partial_t^\alpha u\|_{L^2(0,T)} + \|\partial_t^\alpha a\|_{L^2(0,T)}) \leq C_5(\|\partial_t^\alpha u\|_{L^2(0,T)} + |a|),$$

which is the first inequality in (iv).

(v) By (2.37) and Theorems 2.1 and 2.2, we have

$$\|\partial_t^\alpha u\|_{L^2(0,T)} = \|\partial_t^\alpha (u - u(0))\|_{L^2(0,T)} \sim \|u - u(0)\|_{H_\alpha(0,T)} \sim \|u - u(0)\|_{H^\alpha(0,T)}.$$

Therefore

$$C_6 \|\partial_t^\alpha u\|_{L^2(0,T)} \geq \|u\|_{H^\alpha(0,T)} - \|u(0)\|_{H^\alpha(0,T)}$$

and

$$\|\partial_t^\alpha u\|_{L^2(0,T)} \leq C_6(\|u\|_{H^\alpha(0,T)} + \|u(0)\|_{H^\alpha(0,T)}).$$

Directly, we see that

$$\|u(0)\|_{H^\alpha(0,T)} = |u(0)| \|1\|_{H^\alpha(0,T)} \leq C_7 |u(0)|.$$

Moreover, by the Sobolev embedding for $\frac{1}{2} < \alpha < 1$, we see $|u(0)| \leq C_8 \|u\|_{H^\alpha(0,T)}$. Hence $\|u(0)\|_{H^\alpha(0,T)} \leq C_7 |u(0)| \leq C_9 \|u(0)\|_{H^\alpha(0,T)}$. Thus, the proof of (v) and so the proof of Proposition 2.5 is completed. ∎

It seems that the definition of ∂_t^α in $H^\alpha(0,T)$ with $\frac{1}{2} < \alpha < 1$ may be artificial, but the definition is related to the closure of the classical Caputo derivative operator. In Sect. 2.3, we introduce the closure in $L^2(0,T)$ of the Caputo derivative operator d_t^α with the domain $\mathcal{D}(d_t^\alpha) = {}_0C^1[0, T]$ and we establish Theorem 2.5. Our discussion here is similar. We consider the classical Caputo derivative

$$
{}_0d_t^\alpha u(t) := \frac{1}{\Gamma(1-\alpha)} \int_0^t (t-s)^{-\alpha} \frac{du}{ds}(s)ds, \quad \mathcal{D}({}_0d_t^\alpha) = C^1[0, T].
$$

To avoid confusion, by ${}_0d_t^\alpha$ we denote the classical Caputo fractional derivative with the domain $C^1[0, T]$, and we consider ${}_0d_t^\alpha$ as an operator from $C^1[0, T] \subset L^2(0, T)$ to $L^2(0, T)$.

First we prove:

Lemma 2.6 *For* $\frac{1}{2} \le \alpha < 1$, *the operator* ${}_0d_t^\alpha$ *is closable.*

Proof We have to prove that $w = 0$ if $u_n \in C^1[0, T] = \mathcal{D}({}_0d_t^\alpha)$, $n = 1, 2, 3, \ldots$, $u_n \longrightarrow 0$ in $L^2(0, T)$ and ${}_0d_t^\alpha u_n$ converge to w in $L^2(0, T)$. By $u_n \in C^1[0, T]$, $n = 1, 2, 3, \ldots$, we see that $u_n - u_n(0) \in {}_0C^1[0, T] \subset H_\alpha(0, T)$. Therefore (2.27) implies

$$
{}_0d_t^\alpha u_n = {}_0d_t^\alpha(u_n - u_n(0)) = d_t^\alpha(u_n - u_n(0)) = \partial_t^\alpha(u_n - u_n(0)).
$$

Hence $\partial_t^\alpha(u_n - u_n(0)) \longrightarrow w$ in $L^2(0, T)$, and so Theorem 2.4 implies $u_n - u_n(0)$ is a Cauchy sequence in $H_\alpha(0, T)$, and there exists $v \in H_\alpha(0, T)$ such that

$$
u_n - u_n(0) \longrightarrow v \quad \text{in } H_\alpha(0, T). \tag{2.41}
$$

In particular, $u_n - u_n(0) \longrightarrow v$ in $L^2(0, T)$. Since $u_n \longrightarrow 0$ in $L^2(0, T)$, we have $-u_n(0) \longrightarrow v$ in $L^2(0, T)$ and so $v(t)$ is a constant function as limit of the constant functions: $v(t) = v_0$ for $0 < t < T$ with some constant v_0. By $v \in H_\alpha(0, T)$, we see that $v_0 \in H_\alpha(0, T)$. For $\frac{1}{2} < \alpha < 1$, since $v \in H_\alpha(0, T) \subset C[0, T]$ implies $v(0) = 0$, we obtain $v_0 = 0$. For $\alpha = \frac{1}{2}$, since $v_0 \in H_{\frac{1}{2}}(0, T)$, we have $\int_0^T \frac{|v(t)|^2}{t} dt = v_0^2 \int_0^T \frac{1}{t} dt < \infty$. Consequently, $v_0 = 0$. The proof of Lemma 2.6 is complete. ∎

Now in this section, let $\frac{1}{2} \le \alpha < 1$. By $\widetilde{d_t^\alpha}$, we denote the closure in $L^2(0, T)$ of ${}_0d_t^\alpha$. More precisely, $u \in \mathcal{D}(\widetilde{d_t^\alpha})$ if there exists a sequence $u_n \in C^1[0, T]$, $n = 1, 2, 3, \ldots$ such that $u_n \longrightarrow u$ in $L^2(0, T)$ and ${}_0d_t^\alpha u_n$, $n = 1, 2, 3, \ldots$, are convergent to some $w \in L^2(0, T)$. Then we define $\widetilde{d_t^\alpha} u = w$.

Finally, we prove:

Theorem 2.6

(i)

$$
\mathcal{D}(\widetilde{d_t^\alpha}) = H^\alpha(0, T) \quad if \frac{1}{2} < \alpha < 1
$$

and

$$H_{\frac{1}{2}}(0, T) \subset \mathcal{D}(\widetilde{d_t^\alpha}) \subset H^{\frac{1}{2}}(0, T).$$

(ii) Let $0 < \alpha < \frac{1}{2}$, and let ∂_t^α be defined by (2.37) for all $u \in H^\alpha(0, T)$. Then

$$\widetilde{d_t^\alpha} = \partial_t^\alpha \quad on\ H^\alpha(0, T). \tag{2.42}$$

Proof

First Step We prove

$$\mathcal{D}(\widetilde{d_t^\alpha}) \subset H^\alpha(0, T), \quad \frac{1}{2} \leq \alpha < 1. \tag{2.43}$$

∎

Proof of (2.43) Let $u \in \mathcal{D}(\widetilde{d_t^\alpha})$. The definition of $\widetilde{d_t^\alpha}$ implies that there exists a sequence $u_n \in C^1[0, T]$, $n = 1, 2, 3, \ldots$ such that

$$u_n \longrightarrow u, \qquad {}_0 d_t^\alpha u_n \longrightarrow \widetilde{d_t^\alpha} u \quad \text{in } L^2(0, T). \tag{2.44}$$

Then $u_n - u_n(0) \in {}_0 C^1[0, T] \subset H_\alpha(0, T)$. Lemma 1.3 and (2.27) yield

$$_0 d_t^\alpha u_n = d_t^\alpha(u_n - u_n(0)) = \partial_t^\alpha(u_n - u_n(0)) \longrightarrow \widetilde{d_t^\alpha} u$$

in $L^2(0, T)$. By Theorem 2.4, it follows that $u_n - u_n(0)$, $n = 1, 2, 3, \ldots$, is a Cauchy sequence in $H_\alpha(0, T)$, and so we can find $\widetilde{u} \in H_\alpha(0, T)$ such that

$$u_n - u_n(0) \longrightarrow \widetilde{u} \quad \text{in } H_\alpha(0, T) \subset H^\alpha(0, T). \tag{2.45}$$

In particular, $u_n - u_n(0) \longrightarrow \widetilde{u}$ in $L^2(0, T)$. Therefore, since $u_n \longrightarrow u$ in $L^2(0, T)$, it follows that $-u_n(0) \longrightarrow \widetilde{u} - u$ in $L^2(0, T)$, so that $\widetilde{u} - u$ is a constant function. Hence $-u_n(0) \longrightarrow \widetilde{u} - u$ in $H^\alpha(0, T)$. In terms of (2.45), $u_n \longrightarrow u$ in $H^\alpha(0, T)$. This means that $u \in H^\alpha(0, T)$, and the proof of (2.43) is complete. ∎

Second Step We prove

$$H_\alpha(0, T) \subset \mathcal{D}(\widetilde{d_t^\alpha}) \tag{2.46}$$

and

$$\widetilde{d_t^\alpha} u = \partial_t^\alpha u, \quad u \in H_\alpha(0, T). \tag{2.47}$$

Proof of (2.46) and (2.47) By Lemma 2.2, for any $u \in H_\alpha(0, T)$, we can choose a sequence $u_n \in {}_0C^1[0, T] \subset H_\alpha(0, T)$, $n = 1, 2, 3, \ldots$, such that $u_n \longrightarrow u$ in $H_\alpha(0, T)$. Theorem 2.4 implies that $\partial_t^\alpha u_n \longrightarrow \partial_t^\alpha u$ in $L^2(0, T)$. On the other hand, Lemma 1.3 and (2.27) yield that ${}_0d_t^\alpha u_n = \partial_t^\alpha u_n$, and so ${}_0d_t^\alpha u_n \longrightarrow \partial_t^\alpha u$ in $L^2(0, T)$. Since $u_n \in C^1[0, T]$, by the definition of $\widetilde{d_t^\alpha}$, we obtain $u \in \mathcal{D}(\widetilde{d_t^\alpha})$ and (2.47). ∎

Third Step The inclusions (2.43) and (2.46) imply

$$H_\alpha(0, T) \subset \mathcal{D}(\widetilde{d_t^\alpha}) \subset H^\alpha(0, T). \tag{2.48}$$

To complete the proof of Theorem 2.6, for $\frac{1}{2} < \alpha < 1$, we have to prove

$$H^\alpha(0, T) \subset \mathcal{D}(\widetilde{d_t^\alpha}), \quad \widetilde{d_t^\alpha} u = \partial_t^\alpha(u - u(0)) \quad \text{for } u \in H^\alpha(0, T).$$

Let $u \in H^\alpha(0, T)$ be arbitrary. By $\alpha > \frac{1}{2}$, by the Sobolev embedding we see that $u \in C[0, T]$ and $u - u(0) \in H_\alpha(0, T)$. By Lemma 2.2, we can choose a sequence $u_n \in {}_0C^1[0, T]$, $n = 1, 2, 3, \ldots$, such that

$$u_n \longrightarrow u - u(0) \quad \text{in } H_\alpha(0, T), \tag{2.49}$$

and in particular, $u_n \longrightarrow u - u(0)$ in $L^2(0, T)$. Hence

$$u_n + u(0) \longrightarrow u \quad \text{in } L^2(0, T). \tag{2.50}$$

On the other hand, in terms of (2.49) we see from Theorem 2.4 and (2.27) that

$${}_0d_t^\alpha(u_n + u(0)) = {}_0d_t^\alpha u_n = \partial_t^\alpha u_n \longrightarrow \partial_t^\alpha(u - u(0))$$

in $L^2(0, T)$. With (2.50), we see that $v_n := u_n + u(0) \in C^1[0, T]$ such that $v_n \longrightarrow u$ and ${}_0d_t^\alpha v_n \longrightarrow \partial_t^\alpha(u - u(0))$ in $L^2(0, T)$. Hence the definition of $\widetilde{d_t^\alpha}$ yields $u \in \mathcal{D}(\widetilde{d_t^\alpha})$ and $\widetilde{d_t^\alpha} u = \partial_t^\alpha(u - u(0))$ for $\frac{1}{2} < \alpha < 1$. Thus the proof of Theorem 2.6 is complete. ∎

2.6 Adjoint of the Fractional Derivative in $H_\alpha(0, T)$

In Sects. 2.3 and 2.5, we discussed the two extensions of the classical Caputo derivative by taking the closures. As another method for the extension, we here argue the adjoint operator and show the adjoint equality.

The results in this section are not used in the later parts of the book.

We introduce:

Definition 2.2 The formal adjoint operator $(d_t^\alpha)^*$ to d_t^α is defined by

$$(d_t^\alpha u, \varphi) = (u, (d_t^\alpha)^*\varphi) + \frac{1}{(1-\alpha)\Gamma(1-\alpha)} \left(\int_0^T t^{1-\alpha} \frac{d\varphi}{dt}(t)dt \right) u(0) \qquad (2.51)$$

for $u \in C^1[0, T]$ and $\varphi \in C^1[0, T]$ satisfying $\varphi(T) = 0$.

Then:

Lemma 2.7 *For each $\varphi \in C^1[0, T]$ with $\varphi(T) = 0$, we have*

$$(d_t^\alpha)^*\varphi(t) = \frac{-1}{\Gamma(1-\alpha)} \int_t^T (\eta - t)^{-\alpha} \frac{d\varphi}{d\eta}(\eta)d\eta.$$

Proof Let $u \in C^1[0, T]$. Then, exchanging the orders of the integrals, we have

$$(d_t^\alpha u, \varphi) = \int_0^T \frac{1}{\Gamma(1-\alpha)} \left(\int_0^t (t-s)^{-\alpha} \frac{du}{ds}(s)ds \right) \varphi(t)dt$$

$$= \frac{1}{\Gamma(1-\alpha)} \int_0^T \left(\int_s^T (t-s)^{-\alpha} \varphi(t)dt \right) \frac{du}{ds}(s)ds.$$

On the other hand, by integrating by parts and $\varphi(T) = 0$, we obtain

$$\int_s^T (t-s)^{-\alpha} \varphi(t)dt = \left[\frac{(t-s)^{1-\alpha}}{1-\alpha} \varphi(t) \right]_{t=s}^{t=T} - \frac{1}{1-\alpha} \int_s^T (t-s)^{1-\alpha} \frac{d\varphi}{dt}(t)dt$$

$$= -\frac{1}{1-\alpha} \int_s^T (t-s)^{1-\alpha} \frac{d\varphi}{dt}(t)dt.$$

Therefore again integration by parts yields

$$(d_t^\alpha u, \varphi) = -\frac{1}{(1-\alpha)\Gamma(1-\alpha)} \int_0^T \left(\int_s^T (t-s)^{1-\alpha} \frac{d\varphi}{dt}(t)dt \right) \frac{du}{ds}(s)ds \qquad (2.52)$$

$$= -\frac{1}{(1-\alpha)\Gamma(1-\alpha)} \left[\left(\int_s^T (t-s)^{1-\alpha} \frac{d\varphi}{dt}(t)dt \right) u(s) \right]_{s=0}^{s=T}$$

$$+ \frac{1}{(1-\alpha)\Gamma(1-\alpha)} \int_0^T \frac{d}{ds} \left(\int_s^T (t-s)^{1-\alpha} \frac{d\varphi}{dt}(t)dt \right) u(s)ds$$

$$= \frac{1}{(1-\alpha)\Gamma(1-\alpha)} \left(\int_0^T t^{1-\alpha} \frac{d\varphi}{dt}(t)dt \right) u(0)$$

$$+ \left(u, \frac{1}{(1-\alpha)\Gamma(1-\alpha)} \frac{d}{ds} \int_s^T (t-s)^{1-\alpha} \frac{d\varphi}{dt}(t)dt \right)$$

$$= \left(u, \frac{-1}{\Gamma(1-\alpha)} \int_s^T (t-s)^{-\alpha} \frac{d\varphi}{dt}(t)dt \right) + \frac{1}{(1-\alpha)\Gamma(1-\alpha)} \left(\int_0^T t^{1-\alpha} \frac{d\varphi}{dt}(t)dt \right) u(0)$$

for $u \in C^1[0, T]$ and $\varphi \in C^1[0, T]$ satisfying $\varphi(T) = 0$. The proof of Lemma 2.7 is completed. ∎

Next:

Definition 2.3 For $u \in H_\alpha(0, T)$, we define $\widehat{d_t^\alpha} u = f$ if there exists $f \in L^2(0, T)$ such that

$$(u, (d_t^\alpha)^* \varphi) = (f, \varphi), \quad \varphi \in C_0^1(0, T).$$

The set of such u is defined as the domain of $\widehat{d_t^\alpha}$.

We see that $\widehat{d_t^\alpha} u$ is well-defined, that is, $\widehat{d_t^\alpha} u$ is uniquely determined if $u \in \mathcal{D}(\widehat{d_t^\alpha})$. Indeed, if $\widehat{d_t^\alpha} u = f$ and $\widehat{d_t^\alpha} u = g$, then the definition implies $(f - g, \varphi) = 0$ for all $\varphi \in C_0^1(0, T)$. The density of $C_0^1(0, T)$ in $L^2(0, T)$ implies $f = g$.

Example 2.1 We consider $\widehat{d_t^\alpha} 1$. First by Lemma 2.7, we have

$$(1, (d_t^\alpha)^* \varphi) = \int_0^T \frac{-1}{\Gamma(1-\alpha)} \left(\int_t^T (\eta - t)^{-\alpha} \frac{d\varphi}{d\eta}(\eta)d\eta \right) dt.$$

Exchanging the orders of the integrals and integrating by parts, we obtain

$$(1, (d_t^\alpha)^* \varphi) = \frac{-1}{\Gamma(1-\alpha)} \int_0^T \frac{d\varphi}{d\eta}(\eta) \left(\int_0^\eta (\eta - t)^{-\alpha} dt \right) d\eta$$

$$= -\frac{1}{\Gamma(1-\alpha)} \frac{1}{1-\alpha} \int_0^T \eta^{1-\alpha} \frac{d\varphi}{d\eta}(\eta)d\eta = \frac{1}{\Gamma(1-\alpha)} \int_0^T \eta^{-\alpha} \varphi(\eta)d\eta.$$

By the definition, we see that

$$(\widehat{d_t^\alpha} 1)(t) = \frac{1}{\Gamma(1-\alpha)} t^{-\alpha},$$

which means $(\widehat{d_t^\alpha} 1)(t) = (D_t^\alpha 1)(t) = \frac{1}{\Gamma(1-\alpha)} t^{-\alpha}$, but $d_t^\alpha 1 = 0$.

We can readily prove

$$_0C^1[0, T] \subset \mathcal{D}(\widehat{d_t^\alpha})$$

and the extended $\widehat{d_t^\alpha} u$ coincides with (1.4) for $u \in {}_0C^1[0, T]$.
Indeed, by Lemma 2.1 we note that $_0C^1[0, T] \subset H_\alpha(0, T)$. Setting

$$f(t) = \frac{1}{\Gamma(1 - \alpha)} \int_0^t (t - s)^{-\alpha} \frac{du}{ds}(s) ds$$

and repeating the computations in (2.52), we see that $(d_t^\alpha u, \varphi) = (u, (d_t^\alpha)^* \varphi) = (f, \varphi)$ for all $\varphi \in C_0^1(0, T)$. Thus the proof is complete. ∎

In fact, we can further prove:

Proposition 2.6

$$H_\alpha(0, T) \subset \mathcal{D}(\widehat{d_t^\alpha})$$

and

$$\widehat{d_t^\alpha} u = \partial_t^\alpha u = D_t^\alpha u, \quad u \in H_\alpha(0, T).$$

In the above example, we see that $1 \in \mathcal{D}(\widehat{d_t^\alpha})$, but $1 \notin H_\alpha(0, T)$ for $\frac{1}{2} \le \alpha < 1$. Therefore $H_\alpha(0, T) \subsetneqq \mathcal{D}(\widehat{d_t^\alpha})$ for $\frac{1}{2} \le \alpha < 1$.

Proposition 2.6 generalizes (2.27) with the extension $\widehat{d_t^\alpha}$ and we notice that d_t^α cannot be calculated in general for $u \in H_\alpha(0, T)$.

Example 2.2 Let $0 < \alpha < \frac{1}{2}, \delta > 0$ and $C_0 = \dfrac{\Gamma\left(\delta + \frac{1}{2}\right)}{\Gamma\left(\alpha + \delta + \frac{1}{2}\right)}$. Proposition 2.4 yields $t^{\alpha + \delta - \frac{1}{2}} \in H_\alpha(0, T)$ and Proposition 2.6 implies

$$\widehat{d_t^\alpha}(C_0 t^{\alpha + \delta - \frac{1}{2}}) = \partial_t^\alpha(C_0 t^{\alpha + \delta - \frac{1}{2}}) = D_t^\alpha(C_0 t^{\alpha + \delta - \frac{1}{2}}) = t^{\delta - \frac{1}{2}}.$$

Moreover, we can directly verify the above result by the definition of $\widehat{d_t^\alpha}$. For this, it suffices to prove

$$(C_0 t^{\alpha + \delta - \frac{1}{2}}, (d_t^\alpha)^* \varphi) = (t^{\delta - \frac{1}{2}}, \varphi), \quad \varphi \in C_0^1(0, T).$$

Indeed, using Lemma 2.7, integration by parts and $\int_0^T \left(\int_t^T \cdots d\eta \right) dt = \int_0^T \left(\int_0^\eta \cdots dt \right) d\eta$, we obtain

$$(C_0 t^{\alpha + \delta - \frac{1}{2}}, (d_t^\alpha)^* \varphi) = C_0 \int_0^T t^{\alpha + \delta - \frac{1}{2}} \left(\frac{-1}{\Gamma(1 - \alpha)} \int_t^T (\eta - t)^{-\alpha} \frac{d\varphi}{d\eta}(\eta) d\eta \right) dt$$

$$= \frac{-C_0}{\Gamma(1-\alpha)} \int_0^T \frac{d\varphi}{d\eta}(\eta) \left(\int_0^\eta t^{\alpha+\delta-\frac{1}{2}}(\eta-t)^{-\alpha} dt \right) d\eta = -\frac{1}{\delta+\frac{1}{2}} \int_0^T \eta^{\delta+\frac{1}{2}} \frac{d\varphi}{d\eta}(\eta) d\eta$$

$$= -\frac{1}{\delta+\frac{1}{2}} [\varphi(\eta)\eta^{\delta+\frac{1}{2}}]_{\eta=0}^{\eta=T} + \int_0^T \eta^{\delta-\frac{1}{2}} \varphi(\eta) d\eta = (t^{\delta-\frac{1}{2}}, \varphi).$$

Proof of Proposition 2.6 Let $u \in H_\alpha(0, T)$ be arbitrary. By Lemma 2.2, there exist $u_n \in {_0}C^1[0, T]$, $n = 1, 2, 3, \ldots$, such that $u_n \longrightarrow u$ in $H_\alpha(0, T)$. By Theorem 2.4, we note that $\partial_t^\alpha u_n = d_t^\alpha u_n$, $n = 1, 2, 3, \ldots$. By Theorem 2.2, we have $d_t^\alpha u_n = \partial_t^\alpha u_n \longrightarrow \partial_t^\alpha u$ in $L^2(0, T)$. By using $u_n(0) = 0$, Lemma 2.7, $u_n \in {_0}C^1[0, T]$ and Definition 2.3 imply

$$(d_t^\alpha u_n, \varphi) = (u_n, (d_t^\alpha)^* \varphi), \quad \varphi \in C_0^1(0, T), \, n = 1, 2, 3, \ldots$$

Letting $n \to \infty$, we obtain

$$(u, (d_t^\alpha)^* \varphi) = (\partial_t^\alpha u, \varphi), \quad \varphi \in C_0^1(0, T).$$

Since $\partial_t^\alpha u \in L^2(0, T)$ for $u \in H_\alpha(0, T)$, Definition 2.3 yields that $u \in \mathcal{D}(\widehat{d_t^\alpha})$ and $\widehat{d_t^\alpha} u = \partial_t^\alpha u$. Combining (2.26), we complete the proof of Proposition 2.6. ∎

The adjoint equality (2.51) for $u \in {_0}C^1[0, T]$ is generalized as follows.

Proposition 2.7

$$(\partial_t^\alpha u, \varphi) = (u, (d_t^\alpha)^* \varphi), \quad u \in H_\alpha(0, T), \, \varphi \in C_0^1(0, T). \tag{2.53}$$

Proof For $u \in H_\alpha(0, T)$, by Lemma 2.2, we can choose $u_n \in {_0}C^1[0, T]$, $n \in \mathbb{N}$ such that $u_n \longrightarrow u$ in $H_\alpha(0, T)$. Then, by Lemma 2.7 we have

$$(\partial_t^\alpha u_n, \varphi) = (u_n, (d_t^\alpha)^* \varphi), \quad \varphi \in C_0^1(0, T).$$

Moreover, (2.11) yields that $\partial_t^\alpha u_n \longrightarrow \partial_t^\alpha u$ in $L^2(0, T)$. Therefore letting $n \to \infty$, we complete the proof of Proposition 2.7. ∎

2.7 Laplace Transform of ∂_t^α

For u belonging to a certain class, we can define the Laplace transform $(Lu)(p)$ by

$$(Lu)(p) := \int_0^\infty e^{-pt} u(t) dt$$

for Re $p > p_0$: some constant.

The formulae of the Laplace transforms for fractional derivatives are well-known (e.g., [15, 24]), provided that u satisfies a certain conditions. For example,

$$L(d_t^\alpha u)(p) = p^\alpha (Lu)(p) - p^{\alpha-1} u(0) \qquad (2.54)$$

for Re $p > p_0$: some constant. The formula (2.54) is convenient for solving fractional differential equations. However, formula (2.54) requires some regularity for u in order that $u(0)$ is well-defined. Formula (2.54) does not make sense for all $u \in H^\alpha(0, T)$ with $0 < \alpha < \frac{1}{2}$.

Moreover, such needed regularity should be consistent with the regularity which we can prove for solutions to fractional differential equations. In particular, the regularity for the formula concerning the Laplace transform should be not very strong. Thus on the regularity assumption for the formula like (2.54), we have to make adequate assumptions for u.

In this section, we state the formula of the Laplace transform for the fractional derivative ∂_t^α in $H_\alpha(0, T)$.

We set

$$V_\alpha(0, \infty) := \{u; \ u|_{(0,T)} \in H_\alpha(0, T) \quad \text{for any } T > 0, \qquad (2.55)$$

there exists a constant $C = C_u > 0$ such that $|u(t)| \le Ce^{Ct}$ for $t \ge 0\}$.

Here $u|_{(0,T)}$ denotes the restriction of u to $(0, T)$. Then we can state:

Theorem 2.7 *The Laplace transform $L(\partial_t^\alpha u)(p)$ can be defined for $u \in V_\alpha(0, \infty)$ by*

$$L(\partial_t^\alpha u)(p) = \lim_{T \to \infty} \int_0^T e^{-pt} \partial_t^\alpha u(t) dt, \quad p > C_u$$

and

$$L(\partial_t^\alpha u)(p) = p^\alpha Lu(p), \quad p > C_u.$$

Here $C_u > 0$ in some constant depending on u.

Proof First for $u \in H_\alpha(0, T)$, by Theorem 2.3 (i), we can see that

$$J^{1-\alpha}u \in H_1(0, T) \subset H^1(0, T), \quad J^{1-\alpha}u(0) = 0 \quad \text{in the trace sense} \qquad (2.56)$$

and so

$$D_t^\alpha u = \frac{d}{dt}J^{1-\alpha}u \in L^2(0, T).$$

Theorem 2.4 yields $\partial_t^\alpha u = \dfrac{d}{dt}J^{1-\alpha}u$ for $u \in H_\alpha(0, T)$. Let $T > 0$ be arbitrarily fixed. Then, in terms of (2.56), we integrate by parts to obtain

$$\int_0^T e^{-pt}\partial_t^\alpha u(t)dt = \int_0^T \frac{d}{dt}(J^{1-\alpha}u)(t)e^{-pt}dt$$

$$= \left[J^{1-\alpha}u(t)e^{-pt}\right]_{t=0}^{t=T} + p\int_0^T e^{-pt}J^{1-\alpha}u(t)dt$$

$$= \frac{e^{-pT}}{\Gamma(1-\alpha)}\int_0^T (T-s)^{-\alpha}u(s)ds + \frac{p}{\Gamma(1-\alpha)}\int_0^T e^{-pt}\left(\int_0^t (t-s)^{-\alpha}u(s)ds\right)dt$$

$$=: I_1 + I_2.$$

Since $|u(t)| \le C_0 e^{C_0 t}$ for $t \ge 0$ with some constant $C_0 > 0$, we estimate

$$|I_1| \le Ce^{-pT}\int_0^T (T-s)^{-\alpha}e^{C_0 s}ds = Ce^{-pT}\int_0^T s^{-\alpha}e^{C_0(T-s)}ds$$

$$= Ce^{-(p-C_0)T}\int_0^T s^{-\alpha}e^{-C_0 s}ds \le Ce^{-(p-C_0)T}\int_0^\infty s^{-\alpha}e^{-C_0 s}ds$$

$$= Ce^{-(p-C_0)T}\frac{\Gamma(1-\alpha)}{C_0^{1-\alpha}}.$$

Hence if $p > C_0$, then $\lim_{T\to\infty} I_1 = 0$.

As for I_2, by $\int_0^T \left(\int_0^t \cdots ds\right)dt = \int_0^T \left(\int_s^T \cdots dt\right)ds$, we see

$$I_2 = \frac{p}{\Gamma(1-\alpha)}\int_0^T \left(\int_s^T e^{-pt}(t-s)^{-\alpha}dt\right)u(s)ds$$

$$= \frac{p}{\Gamma(1-\alpha)}\int_0^T \left(\int_0^{T-s} e^{-p\eta}\eta^{-\alpha}d\eta\right)e^{-ps}u(s)ds.$$

For $p > C_0$, since $|u(s)| \le C_0 e^{C_0 s}$ for $s \ge 0$, we have

$$\left\| \int_0^{T-s} e^{-p\eta} \eta^{-\alpha} d\eta \right| e^{-ps} u(s) \right| \le C_0 \left(\int_0^\infty e^{-p\eta} \eta^{-\alpha} d\eta \right) e^{-(p-C_0)s}$$

for all $s > 0$ and $T > 0$, and the Lebesgue convergence theorem yields

$$\lim_{T \to \infty} I_2 = \frac{p}{\Gamma(1-\alpha)} \int_0^\infty \left(\int_0^\infty e^{-p\eta} \eta^{-\alpha} d\eta \right) e^{-ps} u(s) ds$$

$$= \frac{p}{\Gamma(1-\alpha)} \frac{\Gamma(1-\alpha)}{p^{1-\alpha}} \int_0^\infty e^{-ps} u(s) ds = p^\alpha (Lu)(p)$$

for $p > C_0$. Thus the proof of Theorem 2.7 is complete. ∎

Chapter 3
Fractional Ordinary Differential Equations

3.1 Examples

First we consider simple fractional ordinary differential equations:

$$D_t^\alpha u(t) = -\lambda u(t) + f(t), \quad 0 < t < T, \tag{3.1}$$

$$d_t^\alpha u(t) = -\lambda u(t) + f(t), \quad 0 < t < T. \tag{3.2}$$

We note that (3.1) and (3.2), etc. are considered pointwise.

It is known that as a well-posed problem, we have to attach some initial condition, so that we usually discuss the following initial value problems:

$$\begin{cases} D_t^\alpha u(t) = -\lambda u(t) + f(t), \quad 0 < t < T, \\ \lim_{t \downarrow 0} J^{1-\alpha} u(t) = a, \end{cases} \tag{3.3}$$

and

$$\begin{cases} d_t^\alpha u(t) = -\lambda u(t) + f(t), \quad 0 < t < T, \\ u(0) = a. \end{cases} \tag{3.4}$$

It is proved that there exists a unique solution to (3.3) and (3.4) respectively, and the solutions are given by the following formulae (Kilbas et al. [15], Podlubny [24]).

For (3.3)

$$u(t) = at^{\alpha-1} E_{\alpha,\alpha}(-\lambda t^\alpha) + \int_0^t (t-s)^{\alpha-1} E_{\alpha,\alpha}(-\lambda(t-s)^\alpha) f(s) ds, \quad 0 < t < T. \tag{3.5}$$

A. Kubica et al., *Time-Fractional Differential Equations*, SpringerBriefs in Mathematics, https://doi.org/10.1007/978-981-15-9066-5_3

For (3.4)

$$u(t) = aE_{\alpha,1}(-\lambda t^{\alpha}) + \int_0^t (t-s)^{\alpha-1} E_{\alpha,\alpha}(-\lambda(t-s)^{\alpha}) f(s)ds, \quad 0 < t < T.$$

$$(3.6)$$

Here we recall that $E_{\alpha,\beta}(z)$ with $\alpha, \beta > 0$ are defined by (2.28).

Here we should expect that $u(t)$ given by (3.5) and (3.6) are verified to pointwisely satisfy (3.3) and (3.4) respectively. Moreover, in general, for $f \in L^2(0, T)$, we do not know whether the system

$$\begin{cases} D_t^{\alpha} u(t) = -\lambda u(t) + f(t), & 0 < t < T, \\ u(0) = a \end{cases}$$

possesses a unique solution.

Even for a simple initial value problem (3.4) for a fractional ordinary differential equation, there are very few existing works which systematically establish the uniqueness and the existence of solutions in relevant Sobolev spaces with $f \in L^2(0, T)$, which is a natural choice of a function space for f. The purpose of this chapter is to establish the unique existence of solutions to initial value problems for fractional ordinary differential equations.

When we will treat the problem pointwise, for example, not in $L^2(0, T)$, we have to formulate the initial condition in (3.4) and interpret $d_t^{\alpha} u$ for $f \in L^2(0, T)$. The latter is not direct as is seen in Sect. 2.1 of Chap. 2. In this book, on the basis of the fractional derivative ∂_t^{α} defined in Chap. 2, we will discuss exclusively in the Sobolev spaces $H_{\alpha}(0, T)$ with $0 < \alpha < 1$.

In Sect. 3.4, we will return to formulation (3.4) and formula (3.6), and discuss some details.

3.2　Fundamental Inequalities: Coercivity

In the case of $\alpha = 1$, it is trivial that

$$\int_0^t \frac{du}{ds}(s)u(s)ds = \frac{1}{2}|u(t)|^2 - \frac{1}{2}|u(0)|^2, \quad 0 < t < T \qquad (3.7)$$

for $u \in C^1[0, T]$. An inequality of the type (3.7) is useful for proving the uniqueness of solutions. Thanks to (3.7), we can easily verify that if $u \in H^1(0, T)$ satisfies

$$\frac{du}{dt}(t) = \lambda u, \quad 0 < t < T, \qquad u(0) = 0,$$

then $u = 0$ in $(0, T)$. Indeed, multiplying this equation by u and applying (3.7) and the Gronwall inequality, we obtain $u(t) = 0$ for $0 < t < T$.

In this section, we discuss inequalities corresponding to (3.7) within the Sobolev space $H_\alpha(0, T)$.

We prove:

Theorem 3.1

(i)

$$\int_0^T u(t)d_t^\alpha u(t)dt \geq \frac{1}{2\Gamma(1-\alpha)}T^{-\alpha}\|u\|_{L^2(0,T)}^2 - \frac{1}{2\Gamma(2-\alpha)}T^{1-\alpha}|u(0)|^2$$

(3.8)

for all $u \in W^{1,1}(0, T)$.

(ii)

$$\int_0^T u(t)\partial_t^\alpha u(t)dt \geq \frac{1}{2\Gamma(1-\alpha)}T^{-\alpha}\|u\|_{L^2(0,T)}^2$$

(3.9)

for all $u \in H_\alpha(0, T)$.

Theorem 3.2

(i)

$$\frac{1}{\Gamma(\alpha)}\int_0^t (t-s)^{\alpha-1}u(s)d_s^\alpha u(s)ds \geq \frac{1}{2}(|u(t)|^2 - |u(0)|^2), \quad 0 < t < T$$

(3.10)

for all $u \in W^{1,1}(0, T)$.

(ii)

$$\frac{1}{\Gamma(\alpha)}\int_0^t (t-s)^{\alpha-1}u(s)\partial_s^\alpha u(s)ds \geq \frac{1}{2}|u(t)|^2, \quad almost\ all\ t \in (0, T)$$

(3.11)

for all $u \in H_\alpha(0, T)$.

Similar inequalities have been known under a certain regularity condition on $u = u(t)$ and we refer for example, to Alsaedi et al. [3], Bajlekova [4], but we need the inequalities in $H_\alpha(0, T)$ for generalized fractional derivative ∂_t^α.

Henceforth C, $C_k > 0$ denote generic constants which are dependent of T and α, but independent of choices of a, f, $u(t)$, etc.

Proof of Theorem 3.1 *(i)* First we note

$$\overline{C^1[0, T]}^{W^{1,1}(0,T)} = W^{1,1}(0, T).$$

(3.12)

Although this can be proved by a standard method based on the mollifier, we will give the proof at the end of Theorem 3.2.

In view of (3.12), it suffices to prove (3.8) for $u \in C^1[0, T]$. Indeed, we assume that (3.8) is already proved for $u \in C^1[0, T]$. For arbitrary $u \in W^{1,1}(0, T)$, by (3.12) we can choose a sequence $u_n \in C^1[0, T]$ such that $u_n \longrightarrow u$ in $W^{1,1}(0, T)$. Then

$$\int_0^T u_n(t) d_t^\alpha u_n(t) dt$$

$$\geq \frac{T^{-\alpha}}{2\Gamma(1 - \alpha)} \|u_n\|_{L^2(0,T)}^2 - \frac{T^{1-\alpha}}{2\Gamma(2 - \alpha)} |u_n(0)|^2, \quad n \in \mathbb{N}. \tag{3.13}$$

By Lemma 1.3 (ii) we have $d_t^\alpha u_n \longrightarrow d_t^\alpha u$ in $L^1(0, T)$. By the Sobolev embedding $W^{1,1}(0, T) \subset C[0, T]$, the convergence $u_n \longrightarrow u$ in $W^{1,1}(0, T)$ implies $u_n \longrightarrow u$ in $C[0, T]$ and $u_n(0) \longrightarrow u(0)$. Therefore, letting $n \to \infty$ in (3.13), we obtain (3.8) for each $u \in W^{1,1}(0, T)$.

Now we prove (3.8) for $u \in C^1[0, T]$. Henceforth for simplicity, we set

$$g(t) = \frac{1}{\Gamma(1 - \alpha)} t^{-\alpha}, \quad 0 < \alpha < 1, \ t > 0.$$

Then

$$d_t^\alpha u(t) = \int_0^t g(t - s) \frac{du}{ds}(s) ds.$$

We have

$$u(t) d_t^\alpha u(t) = u(t) \int_0^t g(t - s) \frac{du}{ds}(s) ds \tag{3.14}$$

$$= -\int_0^t g(t - s) \frac{du}{ds}(s)(u(s) - u(t)) ds + \int_0^t g(t - s) u(s) \frac{du}{ds}(s) ds$$

$$=: S_1 + S_2.$$

By integration by parts and $\frac{dg}{ds}(s) \leq 0$ for $0 < s < T$, we see

$$S_1 = -\int_0^t g(t - s) \left(\frac{d}{ds}(u(s) - u(t)) \right) (u(s) - u(t)) ds \tag{3.15}$$

$$= -\frac{1}{2} \int_0^t g(t - s) \frac{d}{ds} \left(|u(s) - u(t)|^2 \right) ds$$

$$= -\frac{1}{2} \left[g(t - s)|u(s) - u(t)|^2 \right]_{s=0}^{s=t} - \frac{1}{2} \int_0^t \frac{dg}{d\xi}(t - s)|u(s) - u(t)|^2 ds$$

$$\geq -\frac{1}{2} \left[g(t - s)|u(s) - u(t)|^2 \right]_{s=0}^{s=t} = \frac{1}{2} g(t)|u(t) - u(0)|^2 \geq 0.$$

For the last inequality, we used

$$\lim_{s \to t} g(t-s)|u(s) - u(t)|^2 \le C \lim_{s \to t} (t-s)^{-\alpha}|t-s|^2 = 0$$

by $u \in C^1[0, T]$. Moreover, we have

$$S_2 = \int_0^t g(s)u(t-s)\frac{du}{d\xi}(t-s)ds = \frac{1}{2}\int_0^t g(s)\frac{d}{dt}((|u(t-s)|^2)ds = \frac{1}{2}\partial_t^\alpha|u(t)|^2.$$

$$(3.16)$$

Therefore Lemma 1.3 (i) yiellds

$$S_2 = \frac{1}{2}\frac{d}{dt}\int_0^t g(s)|u(t-s)|^2 ds - \frac{1}{2}g(t)|u(0)|^2.$$

Hence, since $g(T-s) \ge \frac{T^{-\alpha}}{\Gamma(1-\alpha)}$ for $0 < s < T$, we obtain

$$\int_0^T u(t)d_t^\alpha u(t)dt = \int_0^T S_1 dt + \int_0^T S_2 dt \ge \int_0^T S_2(t)dt$$

$$\ge \frac{1}{2}\int_0^T g(s)|u(T-s)|^2 ds - \frac{1}{2}\left(\int_0^T g(t)dt\right)|u(0)|^2$$

$$= \frac{1}{2}\int_0^T g(T-s)|u(s)|^2 ds - \frac{1}{2}\frac{T^{1-\alpha}|u(0)|^2}{(1-\alpha)\Gamma(1-\alpha)}$$

$$\ge \frac{1}{2}\frac{T^{-\alpha}}{\Gamma(1-\alpha)}\|u\|_{L^2(0,T)}^2 - \frac{1}{2}\frac{T^{1-\alpha}|u(0)|^2}{\Gamma(2-\alpha)}$$

for $u \in C^1[0, T]$. Thus the proof of (3.8) for $u \in C^1[0, T]$ is completed, and the proof of (i) is completed. ∎

Proof of Theorem 3.1 (ii) Since $\overline{{}_0C^1[0, T]}^{H_\alpha(0,T)} = H_\alpha(0, T)$ by Lemma 2.2, it suffices to prove (3.9) for $u \in {}_0C^1[0, T]$. Indeed, for each $u \in H_\alpha(0, T)$, there exists a sequence $u_n \in {}_0C^1[0, T]$, $n \in \mathbb{N}$ such that $u_n \longrightarrow u$ in $H_\alpha(0, T)$. Then $\partial_t^\alpha u_n \longrightarrow \partial_t^\alpha u$ in $L^2(0, T)$ by Theorem 2.2. Therefore by the passage to the limit, we obtain (3.9) for each $u \in H_\alpha(0, T)$.

The proof of (3.9) for $u \in {}_0C^1[0, T]$ is done as follows. For $u \in {}_0C^1[0, T]$, equalities (2.27) and Lemma 1.3 (i) yield

$$\partial_t^\alpha u = d_t^\alpha u.$$

Thus we can repeat the proof of (i) to prove (3.9) for $u \in {}_0C^1[0, T]$. Thus the proof of Theorem 3.1 is completed. ∎

Proof of Theorem 3.2 (i) First let $u \in C^1[0, T]$. Then (3.14) and (3.15) imply

$$u(s)d_s^\alpha u(s) := S_1 + S_2 \geq S_2 = \frac{1}{2}\frac{1}{\Gamma(1-\alpha)}\int_0^t (t-s)^{-\alpha}\frac{d}{ds}(|u(s)|^2)ds$$

$$= \frac{1}{2}d_s^\alpha(|u(s)|^2), \quad 0 < s < T.$$

Hence, since $d_s^\alpha c = 0$ for a constant c, we see that

$$u(s)d_s^\alpha u(s) \geq \frac{1}{2}d_s^\alpha(|u(s)|^2) = \frac{1}{2}d_s^\alpha(|u(s)|^2 - |u(0)|^2), \quad 0 < s < T$$

for $u \in C^1[0, T]$. By Theorem 2.4 and $|u(s)|^2 - |u(0)|^2 \in {}_0C^1[0, T]$, we have

$$d_s^\alpha(|u(s)|^2 - |u(0)|^2) = \frac{d}{dt}J^{1-\alpha}(|u(s)|^2 - |u(0)|^2) = J^{-\alpha}(|u(s)|^2 - |u(0)|^2).$$

Therefore

$$u(s)d_s^\alpha u(s) \geq \frac{1}{2}J^{-\alpha}(|u(s)|^2 - |u(0)|^2), \quad 0 < s < T, \ u \in C^1[0, T]. \qquad (3.17)$$

Since

$$J^\alpha u \geq J^\alpha v \quad \text{in } (0, T) \text{ if } u \geq v \text{ in } (0, T),$$

which can be verified by $(t-s)^{-\alpha} > 0$ for $0 < s < t$, by (3.17) we have

$$J^\alpha(ud_t^\alpha u)(t) \geq \frac{1}{2}J^\alpha(J^{-\alpha}(|u|^2 - |u(0)|^2))(t)$$

$$= \frac{1}{2}(|u(t)|^2 - |u(0)|^2), \quad 0 < t < T,$$

and so

$$\frac{1}{\Gamma(\alpha)}\int_0^t (t-s)^{\alpha-1}(d_s^\alpha u(s))u(s)ds \geq \frac{1}{2}(|u(t)|^2 - |u(0)|^2), \quad 0 < t < T$$

for $u \in C^1[0, T]$.

Next, let $u \in W^{1,1}(0, T)$. Then by (3.12) we can choose $u_k \in C^1[0, T]$, $k \in \mathbb{N}$ such that $u_k \longrightarrow u$ in $W^{1,1}(0, T)$. Then $d_t^\alpha u_k \longrightarrow d_t^\alpha u$ in $L^1(0, T)$ by Lemma 1.3. Moreover,

$$\frac{1}{\Gamma(\alpha)}\int_0^t (t-s)^{\alpha-1}u_k(s)d_s^\alpha u_k(s)ds \geq \frac{1}{2}(|u_k(t)|^2 - |u_k(0)|^2), \quad k \in \mathbb{N}. \qquad (3.18)$$

The Young inequality for the convolution (Lemma A.1) implies

$$\left\| \frac{1}{\Gamma(\alpha)} \int_0^t (t-s)^{\alpha-1} u_k(s) d_s^\alpha u_k(s) ds - \frac{1}{\Gamma(\alpha)} \int_0^t (t-s)^{\alpha-1} u(s) d_s^\alpha u(s) ds \right\|_{L^1(0,T)}$$

$$\leq C \int_0^T (t-s)^{\alpha-1} ds \| u_k d_t^\alpha u_k - u d_t^\alpha u \|_{L^1(0,T)}$$

$$\leq C T^\alpha \| u_k d_t^\alpha u_k - u d_t^\alpha u \|_{L^1(0,T)} \longrightarrow 0$$

as $k \to \infty$ by $d_t^\alpha u_k \longrightarrow d_t^\alpha u$ in $L^1(0,T)$ and $u_k \longrightarrow u$ in $W^{1,1}(0,T) \subset C[0,T]$. Therefore we can choose a subsequence $k' \in \mathbb{N}$ tending to ∞ such that

$$\frac{1}{\Gamma(\alpha)} \int_0^t (t-s)^{\alpha-1} u_{k'}(s) d_s^\alpha u_{k'}(s) ds \longrightarrow \frac{1}{\Gamma(\alpha)} \int_0^t (t-s)^{\alpha-1} u(s) d_s^\alpha u(s) ds$$

for almost all $t \in (0,T)$ as $k' \to \infty$. Moreover, since $u_{k'} \longrightarrow u$ in $L^1(0,T)$ as $k' \to \infty$, with suitable subsequence $k'' \in \mathbb{N}$ of k', we have $\lim_{k'' \to \infty} u_{k''}(t) = u(t)$ for almost all $t \in (0,T)$. Letting $k'' \to \infty$ in (3.18) with $k = k''$, we complete the proof of Theorem 3.2 (i). ∎

Proof of Theorem 3.2 (ii) Let $u \in H_\alpha(0,T)$. By Lemma 2.2, we can choose a sequence $u_n \in {}_0C^1[0,T]$ such that $u_n \longrightarrow u$ in $H_\alpha(0,T)$ and $\partial_t^\alpha u_n \longrightarrow \partial_t^\alpha u$ in $L^2(0,T)$. Moreover, $\partial_t^\alpha u_n = d_t^\alpha u_n$ by (2.27). Theorem 3.2 (i) yields

$$\frac{1}{\Gamma(\alpha)} \int_0^t (t-s)^{\alpha-1} u_n(s) \partial_t^\alpha u_n(s) ds \geq \frac{1}{2} |u_n(t)|^2, \quad 0 < t < T.$$

Letting $n \to \infty$, we complete the proof of (ii). ∎

Proof of (3.12) Let $u \in W^{1,1}(0,T)$. We consider the mollifier defined as follows. By the extension theorem, we can take $\widetilde{u} \in W^{1,1}(\mathbb{R})$ such that supp $\widetilde{u}$ is compact and $\widetilde{u}(t) = u(t)$ for $0 \leq t \leq T$. Let $\chi \in C_0^\infty(\mathbb{R})$ satisfy $\chi \geq 0$ in $\mathbb{R}$, supp $\chi \subset \{t; |t| \leq 1\}$ and $\int_{-\infty}^\infty \chi(t) dt = 1$. For $\varepsilon > 0$, we set

$$\chi_\varepsilon(t) = \frac{1}{\varepsilon} \chi \left(\frac{t}{\varepsilon} \right), \quad u_\varepsilon(t) = \int_{-\infty}^\infty \chi_\varepsilon(t-s) u(s) ds \tag{3.19}$$

for $u \in L_{loc}^1(\mathbb{R})$. Then defining v_ε by (3.19) for $v \in W^{1,1}(\mathbb{R})$, we see (e.g., Adams [2]) that $v_\varepsilon \in C_0^\infty(\mathbb{R})$ and $v_\varepsilon \longrightarrow v$ in $W^{1,1}(\mathbb{R})$. We set

$$\widetilde{u}_\varepsilon(t) = \int_{-\infty}^\infty \chi_\varepsilon(t-s) \widetilde{u}(s) ds.$$

Since supp $\tilde{u}$ is compact, we see that supp u_ε is compact. Therefore $\tilde{u}_\varepsilon \in C_0^\infty(\mathbb{R})$ and $\tilde{u}_\varepsilon \longrightarrow \tilde{u}$ in $W^{1,1}(\mathbb{R})$, that is, $\tilde{u}_\varepsilon \longrightarrow u$ in $W^{1,1}(0, T)$. Since $\tilde{u}_\varepsilon|_{[0,T]} \in C^1[0, T]$. we have proved (3.12).   ∎

Let $\Omega \subset \mathbb{R}^n$ be a bounded domain wih boundary $\partial\Omega$ of C^2-class, and let $x = (x_1, ..., x_n) \in \mathbb{R}^n$.

Henceforth we regard $u = u(x, t)$ as a mapping from $t \in (0, T)$ to a function $u(\cdot, t)$ in x and for example, we introduce a function space:

$$L^2(0, T; L^2(\Omega)) := \left\{ u = u(x, t); \int_0^T \|u(\cdot, t)\|_{L^2(\Omega)}^2 dt < \infty \right\}.$$

We can similarly define $C([0, T]; L^2(\Omega))$, $C^2([0, T]; H^1(\Omega))$, etc.

Now we formulate Theorems 3.1 and 3.2 in $L^2(0, T; L^2(\Omega))$. Henceforth for $H_0^1(\Omega) \subset L^2(\Omega)$, identifying the dual of $L^2(\Omega)'$ of $L^2(\Omega)$ with itself, we define $H^{-1}(\Omega) = (H_1^0(\Omega))'$. Then $H_0^1(\Omega) \subset L^2(\Omega) \subset H^{-1}(\Omega)$ algebraically and topologically (e.g., Brezis [5]). By $_{H^{-1}(\Omega)}\langle\cdot, \cdot\rangle_{H_0^1(\Omega)}$ we denote the duality pairing.

Theorem 3.3

(i)

$$\int_0^T (d_t^\alpha u(\cdot, t), u(\cdot, t))_{L^2(\Omega)} dt \geq \frac{T^{-\alpha}}{2\Gamma(1-\alpha)} \|u\|_{L^2(0,T;L^2(\Omega))}^2 - \frac{T^{1-\alpha}}{2\Gamma(2-\alpha)} \|u(\cdot, 0)\|_{L^2(\Omega)}^2$$

for $u \in W^{1,1}(0, T; L^2(\Omega))$.

(ii)

$$\int_0^T {}_{H^{-1}(\Omega)}\langle\partial_t^\alpha u(\cdot, t), u(\cdot, t)\rangle_{H_0^1(\Omega)} dt \geq \frac{T^{-\alpha}}{2\Gamma(1-\alpha)} \|u\|_{L^2(0,T;L^2(\Omega))}^2$$

for $u \in L^2(0, T; H_0^1(\Omega)) \cap H_\alpha(0, T; H^{-1}(\Omega))$.

Theorem 3.4

(i)

$$\frac{1}{\Gamma(\alpha)} \int_0^t (t-s)^{\alpha-1} (d_s^\alpha u(\cdot, s), u(\cdot, s))_{L^2(\Omega)} ds \geq \frac{1}{2}(\|u(\cdot, t)\|_{L^2(\Omega)}^2 - \|u(\cdot, 0)\|_{L^2(\Omega)}^2)$$

for each $u \in W^{1,1}(0, T; L^2(\Omega))$.

(ii)

$$\frac{1}{\Gamma(\alpha)} \int_0^t {}_{H^{-1}(\Omega)}\langle\partial_s^\alpha u(\cdot, s), u(\cdot, s)\rangle_{H_0^1(\Omega)}(t-s)^{\alpha-1} ds \geq \frac{1}{2}\|u(\cdot, t)\|_{L^2(\Omega)}^2$$

for each $u \in L^2(0, T; H_0^1(\Omega)) \cap H_\alpha(0, T; H^{-1}(\Omega))$.

Proof In terms of the mollifier, we approximate u by smooth functions u_ε, $\varepsilon > 0$ as follows. Let $\theta \in C_0^\infty(\mathbb{R}^n)$ satisfy $\theta \geq 0$, supp $\theta \subset \{x \in \mathbb{R}^n; |x| \leq 1\}$, and $\int_{\mathbb{R}^n} \theta(x)dx = 1$. For $\varepsilon > 0$, we set

$$\theta_\varepsilon(x) := \frac{1}{\varepsilon^n}\theta\left(\frac{x}{\varepsilon}\right), \quad u_\varepsilon(x,t) := \int_{\mathbb{R}^n} u(y,t)\theta_\varepsilon(x-y)dy$$

(e.g., Adams [2]). Let $u \in L^2(0,T; H_0^1(\Omega)) \cap H_\alpha(0,T; H^{-1}(\Omega))$. Then $u_\varepsilon \in L^2(0,T; C(\overline{\Omega})) \cap H_\alpha(0,T; C(\overline{\Omega}))$, and $\partial_t^\alpha u_\varepsilon \longrightarrow \partial_t^\alpha u$, $\nabla u_\varepsilon \longrightarrow \nabla u$ in $L^2(0,T; L^2(\Omega))$ as $\varepsilon \to \infty$. Moreover, if $u(\cdot,0) \in L^2(\Omega)$ and $\frac{1}{2} < \alpha < 1$, then we see that $u_\varepsilon(\cdot,0) \longrightarrow u(\cdot,0)$ in $L^2(\Omega)$ as $\varepsilon \to \infty$. Therefore, applying Theorem 3.2 to u_ε, $\varepsilon > 0$ and letting $\varepsilon \to 0$, we complete the proof of (i) and (ii) of Theorem 3.4. Theorem 3.3 can be proved similarly by means of u_ε. ∎

3.3 Well-Posedness for Single Linear Fractional Ordinary Differential Equations

In Chap. 2, we define ∂_t^α in $H_\alpha(0,T)$. For example, the domain $\mathcal{D}(\partial_t^\alpha)$ requires that the element vanishes at $t = 0$ if $\frac{1}{2} < \alpha < 1$. Thus in (3.4), we have to understand that $u \in \mathcal{D}(\partial_t^\alpha)$, and $u(0) = 0$ in the sense of the trace if $\frac{1}{2} < \alpha < 1$. In order to attach non-zero value to $u(0)$ in some sense not only for $\frac{1}{2} < \alpha < 1$ but also for $0 < \alpha \leq \frac{1}{2}$, we interpret $u(0) = a$ as $u - a \in H_\alpha(0,T)$. Thus in this section, we discuss an initial value problem for a linear fractional ordinary differential equation:

$$\begin{cases} \partial_t^\alpha(u - a) = -\lambda u + f(t), & 0 < t < T, \\ u - a \in H_\alpha(0,T). \end{cases} \tag{3.20}$$

We note that if $\frac{1}{2} < \alpha < 1$, then $u - a \in H_\alpha(0,T)$ yields $u(0) = a$, which can justify the initial condition in the pointwise sense. In view of Proposition 2.5 (i), we can rewrite (3.20) as in the following lemma.

Lemma 3.1 *In the sense of (2.37), the system (3.20) is equivalent to*

$$\begin{cases} \partial_t^\alpha u = -\lambda u + \frac{t^{-\alpha}}{\Gamma(1-\alpha)}a + f(t), & 0 < t < T, \\ u \in H^\alpha(0,T), & \text{if } 0 < \alpha < \frac{1}{2}, \end{cases} \tag{3.21}$$

and

$$\begin{cases} \partial_t^\alpha u = -\lambda u + f(t), & 0 < t < T, \\ u(0) = a. \quad u \in H^\alpha(0,T), & \text{if } \frac{1}{2} < \alpha < 1. \end{cases} \tag{3.22}$$

In the case of $\frac{1}{2} < \alpha < 1$, the initial condition $u(0) = a$ is understood as $\lim_{t\to 0} u(t) = a$, because $H^\alpha(0,T) \subset C[0,T]$.

Proof Let $0 < \alpha < \frac{1}{2}$. Then $H^\alpha(0, T) = H_\alpha(0, T)$ and $u, a \in H_\alpha(0, T)$. Therefore

$$\partial_t^\alpha(u - a) = \partial_t^\alpha u - \partial_t^\alpha a = \partial_t^\alpha u - \frac{t^{-\alpha}}{\Gamma(1 - \alpha)} a.$$

Hence the equivalence between (3.20) and (3.21) is verified. Next, let $\frac{1}{2} < \alpha < 1$. Assume that u satisfies (3.20). By the Sobolev embedding, we see that $u \in C[0, T]$, and $u(0) = a$. Consequently,

$$\partial_t^\alpha(u - a) = \partial_t^\alpha(u - u(0)) = \partial_t^\alpha u$$

by the definition (2.37) of ∂_t^α, which verifies that a solution to (3.20) satisfies (3.22). On the other hand, assume that u satisfies (3.22). We see that $u(0) = a$ implies that $u - a \in H_\alpha(0, T)$. Therefore the definition of ∂_t^α again means that $\partial_t^\alpha u = \partial_t^\alpha(u - a)$. Thus the proof of the lemma is completed.   ∎

For $\alpha = \frac{1}{2}$, we cannot have the corresponding problem like (3.21) and the formulation (3.21) is not convenient. Thus we mainly consider (3.20) as the initial value problem.

We prove:

Theorem 3.5 *Let* $\lambda, a \in \mathbb{R}$ *be given. There exists a unique solution* u *to (3.20). Moreover, for* $0 < \alpha < 1$, *we can choose a constant* $C > 0$ *such that*

$$\|u - a\|_{H_\alpha(0,T)} \le C(|a| + \|f\|_{L^2(0,T)}) \tag{3.23}$$

and

$$\|u\|_{H^\alpha(0,T)} \le C(|a| + \|f\|_{L^2(0,T)}). \tag{3.24}$$

In the pointwise sense, the unique existence of solutions to initial value problems for fractional ordinary differential equations with D_t^α and ∂_t^α has been well studied (e.g., Kilbas et al. [15], Podlubny [24]), but such pointwise formulations meet difficulty in several cases such as $f \notin L^\infty(0, T)$. Thus we need to formulate the initial condition by (3.20).

Proof of Theorem 3.5 By Theorem 2.4, we can rewrite (3.20) as

$$J^{-\alpha}(u - a) = \lambda u + f(t), \quad u - a \in H_\alpha(0, T),$$

which is equivalent to

$$u(t) = a + \lambda(J^\alpha u)(t) + (J^\alpha f)(t)$$

$$= a + \frac{\lambda}{\Gamma(\alpha)} \int_0^t (t - s)^{\alpha-1} u(s)ds + \frac{1}{\Gamma(\alpha)} \int_0^t (t - s)^{\alpha-1} f(s)ds,$$

$$0 < t < T. \tag{3.25}$$

By Theorem 2.1, we see that $\lambda J^{\alpha} : L^2(0, T) \longrightarrow H_{\alpha}(0, T)$ is bounded and so is a compact operator from $L^2(0, T)$ to itself (e.g., [2]). Now we assume that $u(t) = \lambda(J^{\alpha}u)(t)$ for $0 < t < T$, that is,

$$u(t) = \frac{\lambda}{\Gamma(\alpha)} \int_0^t (t-s)^{\alpha-1} u(s) ds, \quad 0 < t < T.$$

Then

$$|u(t)| \leq C \int_0^t (t-s)^{\alpha-1} |u(s)| ds, \quad 0 < t < T.$$

Applying a generalized Gronwall inequality, we obtain $u = 0$ in $(0, T)$. For completeness, we will prove it as Lemma A.2 in the Appendix. ∎

Consequently, the Fredholm alternative yields that there exists a unique solution $u \in L^2(0, T)$ satisfying (3.25). Moreover, since $u - a = J^{\alpha}(\lambda u + f)$ in $L^2(0, T)$. Theorem 2.1 (i) implies that $u - a \in H_{\alpha}(0, T)$. Thus the proof of the unique existence of u is complete.

Moreover, (3.25) yields

$$|u(t)| \leq |a| + C \int_0^t (t-s)^{\alpha-1} |f(s)| ds + C \int_0^t (t-s)^{\alpha-1} |u(s)| ds, \quad 0 < t < T. \tag{3.26}$$

We set

$$R(t) = |a| + C \int_0^t (t-s)^{\alpha-1} |f(s)| ds.$$

Applying the generalized Gronwall inequality Lemma A.2, we have

$$|u(t)| \leq CR(t) + C \int_0^t (t-s)^{\alpha-1} R(s) ds$$

$$\leq C \left(|a| + \int_0^t (t-s)^{\alpha-1} |f(s)| ds \right) + C \int_0^t (t-s)^{\alpha-1} \left(|a| + \int_0^s (s-\xi)^{\alpha-1} |f(\xi)| d\xi \right) ds \tag{3.27}$$

for $0 \leq t \leq T$. We take the norms in $L^2(0, T)$. The Young inequality Lemma A.1 in the Appendix yields

$$\left\| \int_0^t (t-s)^{\alpha-1} |f(s)| ds \right\|_{L^2(0,T)} \leq \|t^{-\alpha}\|_{L^1(0,T)} \|f\|_{L^2(0,T)} \leq \frac{T^{1-\alpha}}{1-\alpha} \|f\|_{L^2(0,T)}.$$

Moreover,

$$\int_0^t \left(\int_0^s (s-\xi)^{\alpha-1} |f(\xi)| d\xi \right) ds = \int_0^t \left(\int_\xi^t (s-\xi)^{\alpha-1} ds \right) |f(\xi)| d\xi$$

$$= \int_0^t \frac{(t-\xi)^\alpha}{\alpha} |f(\xi)| d\xi,$$

and so

$$\left\| \int_0^t \left(\int_0^s (s-\xi)^{\alpha-1} |f(\xi)| d\xi \right) ds \right\|_{L^2(0,T)} \le \left\| \frac{t^\alpha}{\alpha} \right\|_{L^1(0,T)} \|f\|_{L^2(0,T)}$$

again by the Young inequality. By exchanging the orders of the integrals, we can similarly obtain

$$\int_0^t (t-s)^{\alpha-1} \left(\int_0^s (s-\xi)^{\alpha-1} |f(\xi)| d\xi \right) ds$$

$$= \int_0^t |f(\xi)| \left(\int_\xi^t (t-s)^{\alpha-1} (s-\xi)^{\alpha-1} ds \right) d\xi = \frac{\Gamma(\alpha)^2}{\Gamma(2\alpha)} \int_0^t (t-\xi)^{2\alpha-1} |f(\xi)| d\xi,$$

and

$$\left\| \int_0^t (t-s)^{\alpha-1} \left(\int_0^s (s-\xi)^{\alpha-1} |f(\xi)| d\xi \right) ds \right\|_{L^2(0,T)}$$

$$\le \frac{\Gamma(\alpha)^2}{\Gamma(2\alpha)} \left\| \int_0^t (t-\xi)^{2\alpha-1} |f(\xi)| d\xi \right\|_{L^2(0,T)} \le C \|f\|_{L^2(0,T)}.$$

Consequently, (3.27) implies $\|u\|_{L^2(0,T)} \le C(|a| + \|f\|_{L^2(0,T)})$. Hence the first inequality in (3.20) yields

$$\|\partial_t^\alpha (u-a)\|_{L^2(0,T)} \le C(|\lambda| \|u\|_{L^2(0,T)} + \|f\|_{L^2(0,T)}) \le C(|a| + \|f\|_{L^2(0,T)}).$$

Thus the proof of (3.23) is completed. We can see (3.24) by $\|u-a\|_{H^\alpha(0,T)} \le \|u-a\|_{H_\alpha(0,T)}$ and

$$\|u-a\|_{H^\alpha(0,T)} \ge \|u\|_{H^\alpha(0,T)} - \|a\|_{H^\alpha(0,T)} = \|u\|_{H^\alpha(0,T)} - \sqrt{T}|a|,$$

which can be seen by $\|a\|_{H^\alpha(0,T)} = \|a\|_{L^2(0,T)} = \sqrt{T}|a|$ for constant a.

3.4 Alternative Formulation of Initial Value Problem

In this book, we exclusively formulate an initial value problem by (3.20). On the other hand, in this section, we discuss

$$\begin{cases} d_t^\alpha u = p(t)u + f(t), & 0 < t < T, \\ u(0) = a. \end{cases} \tag{3.28}$$

We recall (1.17):

$$W^{1,1}(0, T) := \left\{ u \in L^1(0, T); \frac{du}{dt} \in L^1(0, T) \right\},$$

with $\|u\|_{W^{1,1}(0,T)} = \|u\|_{L^1(0,T)} + \left\| \frac{du}{dt} \right\|_{L^1(0,T)}$.

Here and henceforth we assume $p \in L^\infty(0, T)$ and $f \in L^2(0, T)$.

In (3.28), if $u \in W^{1,1}(0, T)$, then the initial condition $u(0) = a$ can be immediately justified by $W^{1,1}(0, T) \subset C[0, T]$. Moreover, by $u \in W^{1,1}(0, T)$, we can prove that $d_t^\alpha u \in L^1(0, T)$ exists. In many monographs (e.g., Diethelm [6], Kilbas et al. [15], Podlubny [24]), the initial value problem is formulated by (3.28).

Now we consider the formulation:

$$\begin{cases} \partial_t^\alpha(u - a) = p(t)u + f(t), & 0 < t < T, \\ u - a \in H_\alpha(0, T), \end{cases} \tag{3.29}$$

and compare with (3.28). Then we can prove:

Proposition 3.1

(i) *Let $p \in L^\infty(0, T)$ and $f \in L^2(0, T)$. If $u \in W^{1,1}(0, T)$ satisfies (3.28), then u satisfies (3.29).*

(ii) *Let $p \in C^2[0, T]$ and $f \in W^{1,1}(0, T)$. Then the unique solution u to (3.29) is in $W^{1,1}(0, T)$ and satisfies (3.28).*

In (ii) we note that $d_t^\alpha u$ is defined by (1.4) because $u \in W^{1,1}(0, T)$. The proposition means that under assumptions $f \in W^{1,1}(0, T)$ and $p \in C^2[0, T]$, the formulations (3.28) and (3.29) are equivalent. In (3.29), we note that in general we cannot prove that $u \in W^{1,1}(0, T)$ if $p \in L^\infty(0, T)$ and $f \in L^2(0, T)$.

Proof of Proposition 3.1 (i) Let $u \in W^{1,1}(0, T)$ satisfy (3.28). Then Lemma 1.3 (i) implies that $d_t^\alpha u = d_t^\alpha(u - a) = D_t^\alpha(u - a)$ pointwise. Therefore (3.28) yields

$$D_t^\alpha(u - a) = p(t)u + f(t), \quad 0 < t < T. \tag{3.30}$$

Since $u - a \in {}_0W^{1,1}(0, T)$, applying J^α to (3.30), by Proposition 2.1 and $pu + f \in L^2(0, T)$ we obtain

$$u - a = J^\alpha D_t^\alpha(u - a) = J^\alpha(pu + f) \in \mathcal{R}(J^\alpha) = H_\alpha(0, T).$$

Hence $u - a \in H_\alpha(0, T)$. Therefore (2.26) implies $\partial_t^\alpha(u - a) = D_t^\alpha(u - a)$. Equation (3.30) yields (3.29). Thus we complete the proof of Proposition 3.1 (i). ∎

Proof of Proposition 3.1 (ii) For $f \in W^{1,1}(0, T)$ and $p \in C^2[0, T]$, we can prove that $u \in W^{1,1}(0, T)$ and $u(0) = a$. The proof is done in Theorem 3.6 (iii) later for a more general case, and is postponed. Then Theorem 2.4 implies $\partial_t^\alpha(u - a) = d_t^\alpha(u - a)$ for $u - a \in H_\alpha(0, T)$. By $u - a \in W^{1,1}(0, T)$, the pointwise d_t^α defined by (1.4) yields

$$\partial_t^\alpha(u - a) = d_t^\alpha(u - a)(t) = \frac{1}{\Gamma(1 - \alpha)} \int_0^t (t - s)^{-\alpha} \frac{d}{ds}(u - a)(s)ds$$

$$= \frac{1}{\Gamma(1 - \alpha)} \int_0^t (t - s)^{-\alpha} \frac{du}{ds}(s)ds = d_t^\alpha u(t).$$

Therefore (3.29) implies $d_t^\alpha u(t) = p(t)u + f(t)$ for $0 < t < T$. Since $u \in W^{1,1}(0, T) \subset C[0, T]$, we have the initial condition $u(0) = a$ in the sense of $\lim_{t \to 0} u(t) = a$. Thus u satisfies (3.28) and the proof of (ii) is complete. ∎

Now we discuss a simple case (3.4) and clarify in which sense the solution formula (3.6) should be understood.

In fact, we prove:

Proposition 3.2 *Let $f \in L^2(0, T)$. Then u given by (3.6) satisfies*

$$\begin{cases} \partial_t^\alpha(u - a) = -\lambda u + f(t), & 0 < t < T, \\ u - a \in H_\alpha(0, T). \end{cases} \tag{3.31}$$

The existing references [15, 24] give a representation formula (3.6) for the solution to (3.4), but it can be justified pointwise only if f has certain regularity such as $f \in W^{1,1}(0, T)$. In other words, for $f \in L^2(0, T)$ it is more consistent for us to interpret (3.6) as solution formula for the initial value problem (3.31), not (3.4).

Proof We set

$$u_1(t) = aE_{\alpha,1}(-\lambda t^\alpha), \quad u_2(t) = u_2(f)(t) = \int_0^t (t - s)^{\alpha - 1} E_{\alpha,\alpha}(-\lambda(t - s)^\alpha) f(s)ds.$$

Since $u_1 - a \in {}_0W^{1,1}(0, T) \cap H_\alpha(0, T)$ by Proposition 2.2, Eq. (2.27) yields

$$\partial_t^\alpha(u_1 - a) = d_t^\alpha(u_1 - a) = d_t^\alpha u_1 = -\lambda E_{\alpha,1}(-\lambda t^\alpha) = -\lambda u_1(t).$$

Next we have to prove

$$\begin{cases} \partial_t^\alpha u_2 = -\lambda u_2 + f(t), & 0 < t < T, \\ u_2 \in H_\alpha(0, T) \end{cases}$$

for each $f \in L^2(0, T)$. Proposition 2.3 immediately $u_2 \in H_\alpha(0, T)$. We will verify the first equation in the above. First let $f \in C_0^\infty(0, T)$. Then

$$u_2(f)(t) = \int_0^t (t - s)^{\alpha-1} E_{\alpha,\alpha}(-\lambda(t - s)^\alpha) f(s) ds$$

$$= \int_0^t s^{\alpha-1} E_{\alpha,\alpha}(-\lambda s^\alpha) f(t - s) ds.$$

Hence

$$|u_2(f)(t)| \le C \int_0^t s^{\alpha-1} ds \|f\|_{C[0,T]} = \frac{C}{\alpha} t^\alpha \|f\|_{C[0,T]}.$$

In particular, $u_2(f)(0) = 0$. Moreover, using $f \in C_0^\infty(0, T)$, we can justify:

$$\frac{du_2(f)}{dt}(t) = \int_0^t s^{\alpha-1} E_{\alpha,\alpha}(-\lambda s^\alpha) \frac{df}{dt}(t - s) ds.$$

Therefore

$$\left| \frac{du_2(f)}{dt}(t) \right| \le C \int_0^t s^{\alpha-1} ds \|f\|_{C^1[0,T]} = \frac{C}{\alpha} t^\alpha \|f\|_{C^1[0,T]},$$

so that $u_2(f) \in {}_0W^{1,1}(0, T)$. Consequently, $u_2(f) \in H_\alpha(0, T) \cap {}_0W^{1,1}(0, T)$ for $f \in C_0^\infty(0, T)$. Therefore (2.27) yields

$$\partial_t^\alpha u_2(f)(t) = D_t^\alpha u_2(f)(t) = \frac{1}{\Gamma(1 - \alpha)} \frac{d}{dt} \int_0^t (t - s)^{-\alpha} u_2(f)(s) ds, \quad 0 < t < T.$$

Exchanging the orders of the integrals, we calculate

$$\frac{1}{\Gamma(1 - \alpha)} \frac{d}{dt} \int_0^t (t - s)^{-\alpha} u_2(f)(s) ds$$

$$= \frac{1}{\Gamma(1 - \alpha)} \frac{d}{dt} \left(\int_0^t (t - s)^{-\alpha} \left(\int_0^s (s - \xi)^{\alpha-1} E_{\alpha,\alpha}(-\lambda(s - \xi)^\alpha) f(\xi) d\xi \right) ds \right)$$

$$= \frac{d}{dt} \int_0^t \frac{1}{\Gamma(1 - \alpha)} \left(\int_\xi^t (t - s)^{-\alpha} (s - \xi)^{\alpha-1} E_{\alpha,\alpha}(-\lambda(s - \xi)^\alpha) ds \right) f(\xi) d\xi.$$

On the other hand, using the power series of $E_{\alpha,\alpha}(-\lambda\eta^\alpha)$ for $\eta \ge 0$, we can directly verify

$$\frac{1}{\Gamma(1 - \alpha)} \int_0^t (t - \eta)^{-\alpha} \eta^{\alpha-1} E_{\alpha,\alpha}(-\lambda\eta^\alpha) d\eta = E_{\alpha,1}(-\lambda t^\alpha), \quad t > 0,$$

which means

$$\frac{1}{\Gamma(1-\alpha)} \int_\xi^t (t-s)^{-\alpha}(s-\xi)^{\alpha-1} E_{\alpha,\alpha}(-\lambda(s-\xi)^\alpha)ds$$

$$= \frac{1}{\Gamma(1-\alpha)} \int_0^{t-\xi} ((t-\xi)-\eta)^{-\alpha}\eta^{\alpha-1} E_{\alpha,\alpha}(-\lambda\eta^\alpha)d\eta = E_{\alpha,1}(-\lambda(t-\xi)^\alpha).$$

Consequently,

$$\partial_t^\alpha u_2(f)(t) = \frac{d}{dt} \int_0^t E_{\alpha,1}(-\lambda(t-\xi)^\alpha)f(\xi)d\xi.$$

Since

$$\frac{d}{dt} E_{\alpha,1}(-\lambda(t-\xi)^\alpha) = -\lambda(t-\xi)^{\alpha-1} E_{\alpha,\alpha}(-\lambda(t-\xi)^\alpha)$$

by Lemma 2.5 (ii) with $m = 1$, using $f \in C_0^\infty(0, T)$, we have

$$\partial_t^\alpha u_2(f)(t) = \int_0^t -\lambda(t-\xi)^{\alpha-1} E_{\alpha,\alpha}(-\lambda(t-\xi)^\alpha)f(\xi)d\xi + E_{\alpha,1}(0)f(t),$$

that is,

$$\partial_t^\alpha u_2(f) = -\lambda u_2(f) + f(t), \quad 0 < t < T$$

for $f \in C_0^\infty(0, T)$. Finally let $f \in L^2(0, T)$ be arbitrary. We choose $f_n \in C_0^\infty(0, T)$ such that $f_n \longrightarrow f$ in $L^2(0, T)$ as $n \to \infty$. Then we already proved that $\partial_t^\alpha u_2(f_n) = -\lambda u_2(f_n) + f_n(t)$ for $0 < t < T$. Again Proposition 2.3 implies that $u_2(f_n) \longrightarrow u_2(f)$ in $H_\alpha(0, T)$ and $\partial_t^\alpha u_2(f_n) \longrightarrow \partial_t^\alpha u_2(f)$ in $L^2(0, T)$. Hence, letting $n \to \infty$, we obtain $\partial_t^\alpha u_2(f) = -\lambda u_2(f) + f(t)$ also for $f \in L^2(0, T)$. Thus the proof of Proposition 3.2 is completed. ∎

3.5 Systems of Linear Fractional Ordinary Differential Equations

Let $u(t) = \begin{pmatrix} u_1(t) \\ \vdots \\ u_N(t) \end{pmatrix}$ and $F(t) = \begin{pmatrix} f_1(t) \\ \vdots \\ f_N(t) \end{pmatrix} \in (L^2(0, T))^N$, $a = \begin{pmatrix} a_1 \\ \vdots \\ a_N \end{pmatrix} \in \mathbb{R}^N$

and let $P(t) = (p_{ij}(t))_{1 \le i, j \le N}$ with $p_{ij} \in L^\infty(0, T)$ for $1 \le i, j \le N$ be an $N \times N$ matrix.

In this section, we discuss a system of linear fractional ordinary differential equations:

$$\begin{cases} \partial_t^\alpha(u - a) = P(t)u(t) + F(t), & 0 < t < T, \\ u - a \in (H_\alpha(0, T))^N. \end{cases} \tag{3.32}$$

Similarly to Theorem 3.5, we can prove:

Theorem 3.6

(i) *There exists a unique solution u to (3.32).*

(ii) *There exists a constant $C > 0$ such that*

$$\begin{cases} \|u - a\|_{(H_\alpha(0,T))^N} \le C(|a|_{\mathbb{R}^N} + \|F\|_{(L^2(0,T))^N}), \\ \|u\|_{(H^\alpha(0,T))^N} \le C(|a|_{\mathbb{R}^N} + \|F\|_{(L^2(0,T))^N}). \end{cases} \tag{3.33}$$

(iii) *We further assume that $F \in (W^{1,1}(0, T))^N$ and $P \in (C^2[0, T])^{N \times N}$. Then $u \in (W^{1,1}(0, T))^N$ and $u(0) = a$.*

(iv) *Let $F = 0$ and $P \in (C^2[0, T])^{N \times N}$. Then $u - a \in (W_\alpha(0, T))^N$.*

(v) *For $F \in (C^1[0, T])^N$ and $P \in (C^2[0, T])^{N \times N}$, we have $u - a \in (W_\alpha(0, T))^N$.*

The proof of (v) is the same as (iv) and we omit.

Theorem 3.6, in particular parts (iii) and (v) will play an important role in proving the unique existence of solutions to initial boundary value problems in Chap. 4. When we can assume stronger regularity than $P \in (C^2[0, T])^{N \times N}$, we can simplify the proof but we omit details.

Moreover, we recall that

$$W_\alpha(0, T) = \left\{ v \in W^{1,1}(0, T); t^{1-\alpha} \frac{dv}{dt} \in L^\infty(0, T), v(0) = 0 \right\}.$$

Proof of (i) By Theorem 2.4, we can rewrite (3.32) as

$$\begin{cases} J^{-\alpha}(u - a) = P(t)u(t) + F(t), \\ u - a \in (H_\alpha(0, T))^N, \end{cases} \tag{3.34}$$

and so

$$u(t) = a + J^\alpha F(t) + J^\alpha(Pu)(t)$$

$$= a + J^\alpha F(t) + \frac{1}{\Gamma(\alpha)} \int_0^t (t - s)^{\alpha - 1} P(s)u(s) ds, \quad 0 < t < T. \tag{3.35}$$

By Theorem 2.1 and $P \in (L^\infty(0, T))^{N \times N}$, using the compact embedding $(H_\alpha(0, T))^N \longrightarrow (L^2(0, T))^N$, we verify that $u \longmapsto J^\alpha(Pu)$ is a compact operator from $(L^2(0, T))^N$ to $(L^2(0, T))^N$. Therefore in view of the Fredholm alternative, for the unique existence of solution u to (3.32), it suffices to prove

$$u(t) = J^\alpha (Pu)(t), \quad 0 < t < T, \tag{3.36}$$

implies $u = 0$ for $0 < t < T$.

Setting $|u| = \left(\sum_{j=1}^N u_j^2 \right)^{\frac{1}{2}}$, we have

$$|u(t)| \le C \int_0^t (t-s)^{-\alpha} \| P \|_{(L^\infty (0,T))^{N \times N}} |u(s)| ds, \quad 0 < t < T.$$

Hence, by the same argument as in Theorem 3.5, the generalized Gronwall inequality (Lemma A.2) yields $u = 0$ in $(0, T)$. Thus the proof of (i) is completed. ∎

Proof of (ii) By (3.35), we obtain

$$u(t) = a + \frac{1}{\Gamma(\alpha)} \int_0^t (t-s)^{\alpha-1} F(s) ds + \frac{1}{\Gamma(\alpha)} \int_0^t (t-s)^{\alpha-1} P(s) u(s) ds$$

for $0 < t < T$. Consequently, using $P \in (L^\infty (0, T))^{N \times N}$, we have

$$|u(t)| \le CR(t) + C \int_0^t (t-s)^{\alpha-1} |u(s)| ds, \quad 0 < t < T, \tag{3.37}$$

where we set

$$R(t) = |a| + C \int_0^t (t-s)^{\alpha-1} |F(s)| ds.$$

Thus, arguing in the same way as in Theorem 3.5, we can complete the proof of Theorem 3.6 (ii). ∎

Proof of (iii) We set

$$L_\alpha v := J^\alpha (P(t)v), \quad v \in (L^2(0, T))^N, \quad G(t) = J^\alpha (P(t)a) + J^\alpha F(t), \quad 0 < t < T.$$

Moreover, for $0 < \gamma < 1$ and $m \in \mathbb{N}$, we define

$$H_{m+\gamma}(0, T) = \left\{ v \in H^m(0, T); \ v(0) = \cdots = \frac{d^{m-1} v}{dt^{m-1}}(0) = 0, \ \frac{d^m v}{dt^m} \in H_\gamma(0, T) \right\}.$$

By the definition, we see

$$H_{m+\gamma}(0, T) \subset H^{m+\gamma}(0, T). \tag{3.38}$$

Here $H^{m+\gamma}(0, T)$ is a fractional Sobolev space (e.g., Adams [2]), and the norm in $H^{m+\gamma}(0, T)$ is defined by

$$\|v\|_{H^{m+\gamma}(0,T)} = \left(\|v\|_{L^2(0,T)}^2 + \left\| \frac{d^m v}{dt^m} \right\|_{L^2(0,T)}^2 \right.$$

$$\left. + \int_0^T \int_0^T \left| \frac{d^m v}{dt^m}(t) - \frac{d^m v}{dt^m}(s) \right|^2 \frac{1}{|t-s|^{1+2\gamma}} dt ds \right)^{\frac{1}{2}}. \tag{3.39}$$

Furthermore,

$$\|v\|_{H_{m+\gamma}(0,T)} = \begin{cases} \|v\|_{H^{m+\gamma}(0,T)}, & 0 < \gamma < 1, \ \gamma \neq \frac{1}{2}, \\[2mm] \left(\|v\|_{H^{\frac{1}{2}+m}(0,T)}^2 + \int_0^T \frac{1}{t} \left| \frac{d^m v}{dt^m}(t) \right|^2 dt \right)^{\frac{1}{2}}, & \gamma = \frac{1}{2}. \end{cases} \tag{3.40}$$

We show the two lemmata.

Lemma 3.2

$$J^{\alpha}(W^{1,1}(0, T))^N \subset (W^{1,1}(0, T))^N \quad \text{and} \quad L_{\alpha}(W^{1,1}(0, T))^N \subset (W^{1,1}(0, T))^N$$

for $0 < \alpha < 1$.

Lemma 3.3 $L_{\alpha}(H_{\beta}(0, T))^N \subset (H_{\alpha+\beta}(0, T))^N$ *for $0 < \alpha < 1$ and $0 < \beta \leq 2$.*

At the end of the proof of Theorem 3.6, we give the proofs of the lemmata.
Now we proceed to the proof of (iii). Since $u - a = J^{\alpha}(P(u-a) + Pa) + J^{\alpha} F$, in terms of L_{α}, Eq. (3.34) is rewritten as

$$u - a = L_{\alpha}(u - a) + G \quad \text{in } (0, T). \tag{3.41}$$

By Lemma 1.3 (i) and $Pa, F \in (W^{1,1}(0, T))^N$, we can easily verify

$$G \in (W^{1,1}(0, T))^N. \tag{3.42}$$

As is already proved in parts (i) and (ii), we see that $u - a \in (H_{\alpha}(0, T))^N$. Therefore, setting $v_1 = L_{\alpha}(u - a)$ and $w_1 = G$, Eq. (3.41) and Lemma 3.3 imply $u - a = v_1 + w_1$ with $v_1 \in (H_{2\alpha}(0, T))^N$ and $w_1 \in (W^{1,1}(0, T))^N$. Substituting this into (3.41), we obtain

$$u - a = L_{\alpha} v_1 + (L_{\alpha} w_1 + G).$$

By Lemmata 3.2 and 3.3, we see that $L_{\alpha} v_1 \in (H_{3\alpha}(0, T))^N$ and $L_{\alpha} w_1 \in (W^{1,1}(0, T))^N$. Setting $v_2 = L_{\alpha} v_1$ and $w_2 = L_{\alpha} w_1 + G$, we reach $u - a = v_2 + w_2$ with $v_2 \in (H_{3\alpha}(0, T))^N$ and $w_2 \in (W^{1,1}(0, T))^N$.

By $0 < \alpha < 1$, we can find $k_0 \in \mathbb{N}$ such that $1 \le k_0\alpha < 2$. Indeed, we can choose the minimum $k_0 \in \mathbb{N}$ satisfying $k_0\alpha \ge 1$. Then we have $(k_0 - 1)\alpha < 1$. Hence $k_0\alpha < 1 + \alpha < 2$ by $0 < \alpha < 1$.

The above arguments can be repeated up to k_0, That is,

$$u - a = \tilde{v} + \tilde{w} \quad \text{with } \tilde{v} \in (H_{k_0\alpha}(0, T))^N \subset (W^{1,1}(0, T))^N \text{ and } \tilde{w} \in (W^{1,1}(0, T))^N.$$

Therefore we complete the proof that $u - a \in (W^{1,1}(0, T))^N$.

Finally, in terms of (3.41), we have

$$|(u - a)(t)| \le |L_a(u - a)(t)| + |G(t)| \tag{3.43}$$

and ·

$$|L_a(u - a)(t)| = \left| \frac{1}{\Gamma(\alpha)} \int_0^t (t - s)^{\alpha-1} (P(s)(u(s) - a)) ds \right|$$

$$\le C \|P\|_{(L^\infty(0,T))^{N \times N}} \|u - a\|_{(L^\infty(0,T))^N} t^\alpha$$

in view of $u - a \in (W^{1,1}(0, T))^N \subset (L^\infty(0, T))^N$ by the Sobolev embedding. We can directly verify that $|G(t)| \le Ct^\alpha$. Therefore $\lim_{t \to 0} |u(t) - a| = 0$ by (3.43). Thus the proof of (iii) is complete. ∎

Proof of (iv) First we prove: ∎

Lemma 3.4 $L_\alpha h \in (W_\alpha(0, T))^N$ for $h \in (W_\alpha(0, T))^N$.

Proof of Lemma 3.4 Since $h \in (W_\alpha(0, T))^N$, we have $h(0) = 0$, $h \in (L^\infty(0, T))^N$ and $L_\alpha h(0) = 0$. Therefore Lemma 1.3 (i) yields

$$\frac{d}{dt}(L_\alpha h)(t) = \frac{d}{dt} J^\alpha(P(t)h(t)) = J^\alpha \left(\frac{d}{dt}(P(t)h(t)) \right)$$

$$= J^\alpha \left(\frac{dP(t)}{dt} h(t) \right) + J^\alpha \left(P(t) \frac{dh}{dt}(t) \right).$$

Moreover, we can directly verify that

$$J^\alpha L^\infty(0, T) \subset L^\infty(0, T). \tag{3.44}$$

By $P \in (C^1[0, T])^{N \times N}$, we have $\frac{dP}{dt} h \in (L^\infty(0, T))^N$, so that (3.44) implies

$$J^\alpha \left(h \frac{dP}{dt} \right) \in (L^\infty(0, T))^N.$$

By $h \in (W_\alpha(0, T))^N$, we have

$$\left|\frac{dh}{dt}(t)\right| \leq C_h t^{\alpha-1},$$

and so

$$\left|J^\alpha\left(P(t)\frac{dh}{dt}(t)\right)\right| \leq C\int_0^t (t-s)^{\alpha-1}\left|P(t)\frac{dh}{dt}(s)\right|ds$$

$$\leq CC_h\int_0^t (t-s)^{\alpha-1}s^{\alpha-1}ds \leq C_h t^{2\alpha-1} \leq C_1 t^{\alpha-1}.$$

Therefore

$$\left|\frac{d}{dt}(L_\alpha h(t))\right| \leq C_2 t^{\alpha-1},$$

which means that $L_\alpha h \in (W_\alpha(0, T))^N$. Thus the proof of Lemma 3.4 is complete. ∎

By $F = 0$ we have $G(t) = J^\alpha(P(t)a)$, $0 < t < T$. In terms of $P \in (C^1[0, T])^{N\times N}$ and Lemma 1.3 (i), we see that $G(0) = 0$ and

$$\frac{d}{dt}G(t) = J^\alpha\left(\frac{d}{dt}(P(t)a)\right) = J^\alpha\left(\frac{dP(t)}{dt}a\right) + \frac{P(0)a}{\Gamma(\alpha)}t^{\alpha-1}.$$

By (3.44), we obtain $J^\alpha\left(\frac{dP}{dt}a\right) \in (L^\infty(0, T))^N$ and so $\frac{dG}{dt}t^{1-\alpha} \in (L^\infty(0, T))^N$. Consequently,

$$G \in (W_\alpha(0, T))^N. \tag{3.45}$$

In view of Lemma 3.4, we see that

$$L_\alpha^j G \in (W_\alpha(0, T))^N, \quad j = 0, 1, 2, \ldots. \tag{3.46}$$

We choose $k_1 \in \mathbb{N}$ such that

$$k_1\alpha > \frac{3}{2}, \quad (k_1 - 1)\alpha \leq \frac{3}{2}. \tag{3.47}$$

Similarly to the proof of (iii), in terms of (3.41), we will improve the regularity of $u - a$. First by Lemma 3.3, $u - a \in (H_\alpha(0, T))^N$ yields $L_\alpha(u - a) \in (H_{2\alpha}(0, T))^N$. Consequently, setting $v_1 = L_\alpha(u - a)$ and $w_1 = G$, by (3.41) we see $u - a = v_1 + w_1$ with $v_1 \in (H_{2\alpha}(0, T))^N$ and $w_1 \in (W_\alpha(0, T))^N$. Therefore, continuing this argument, in view of Lemmata 3.3 and 3.4, we obtain $u - a = v_k + w_k$ for $k \in \mathbb{N}$, where $v_k \in (H_{(k+1)\alpha}(0, T))^N$ and $w_k \in (W_\alpha(0, T))^N$ provided that $(k-1)\alpha \le 2$. Hence

$$u - a = \widetilde{V} + \widetilde{W}, \quad \widetilde{V} \in (H_{k_1\alpha}(0, T))^N, \quad \widetilde{W} \in (W_\alpha(0, T))^N.$$

By the Sobolev embedding and (3.47), we see that $\widetilde{V} \in (H_{k_1\alpha}(0, T))^N$ implies

$$\frac{d\widetilde{V}}{dt} \in (H_{k_1\alpha-1}(0, T))^N \subset (L^\infty(0, T))^N,$$

which means that $\widetilde{V} \in (W_\alpha(0, T))^N$. Thus $u - a \in (W_\alpha(0, T))^N$, and the proof of Theorem 3.6 is complete.   ∎

Now we prove Lemmata 3.2 and 3.3.

Proof of Lemma 3.2 The first inclusion was proved as Lemma 1.3 (i).
By $P \in (C^1[0, T])^{N\times N}$, we see that $P(W^{1,1}(0, T))^N) \subset (W^{1,1}(0, T))^N$ and so

$$L_\alpha(W^{1,1}(0, T))^N = J^\alpha(P(W^{1,1}(0, T))^N) \subset J^\alpha(W^{1,1}(0, T))^N.$$

Thus $L_\alpha(W^{1,1}(0, T))^N \subset (W^{1,1}(0, T))^N$ follows and the proof of Lemma 3.2 is complete.   ∎

Proof of Lemma 3.3 In terms of (2.4), (3.39) and (3.40), using $P \in (C^2[0, T])^{N\times N}$, we obtain

$$P(H_\beta(0, T))^N \subset (H_\beta(0, T))^N, \quad 0 < \beta \le 2. \tag{3.48}$$

Indeed, for (3.48) we need the assumption $P \in (C^2[0, T])^{N\times N}$, not $P \in (C^1[0, T])^{N\times N}$.

In order to prove Lemma 3.3, thanks to (3.48), it suffices to prove

$$J^\alpha H_\beta(0, T) \subset H_{\alpha+\beta}(0, T) \tag{3.49}$$

for $0 < \beta \le 2$ and $0 < \alpha < 1$.   ∎

Proof of (3.49) For $0 < \alpha + \beta \le 1$ and $0 < \alpha < 1$, estimate (3.49) is already proved in Theorem 2.3 (i). We have to prove more general cases. Let $\alpha + \beta > 1$. It is sufficient to prove (3.49) for $\beta = 1 + \delta$ with $0 \le \delta \le 1$. Let $v \in H_\beta(0, T)$ be arbitrarily given. Then $\frac{dv}{dt} \in H_\delta(0, T)$. Hence $v \in H_\beta(0, T) = H_{1+\delta}(0, T)$ implies $v \in W^{1,1}(0, T)$ and $v(0) = 0$. Moreover,

$$\frac{d}{dt} J^\alpha v = J^\alpha \frac{dv}{dt}$$

by Lemma 1.3 (i) and so

$$\frac{d}{dt} J^\alpha v \in J^\alpha(H_\delta(0, T)).$$

If $\alpha + \delta \leq 1$, then Theorem 2.3 (i) yields $J^\alpha(H_\delta(0, T)) \subset H_{\alpha+\delta}(0, T)$. By the definition of $H_{\alpha+\delta+1}(0, T)$, we see

$$J^\alpha v \in H_{\alpha+\beta}(0, T). \tag{3.50}$$

Next we assume $1 < \alpha + \delta < 2$. We can represent $\alpha + \delta = 1 + \delta_1$ with some $\delta_1 \in (0, 1)$. By $0 \leq \delta \leq 1$ and Theorem 2.1 (i), we obtain $H_\delta(0, T) = J^\delta L^2(0, T)$ and it follows from $\alpha + \delta = 1 + \delta_1$ and Lemma 1.3 (iv) that

$$J^\alpha H_\delta(0, T) = J^\alpha(J^\delta L^2(0, T)) = J^{\alpha+\delta} L^2(0, T) = J^{\delta_1}(J^1 L^2(0, T)).$$

Since $\frac{d}{dt}(J^{\delta_1}(J^1 w)) = J^{\delta_1} \frac{d(J^1 w)}{dt} = J^{\delta_1} w$ for all $w \in L^2(0, T)$, the definition of $H_{1+\delta_1}(0, T)$ yields $J^{1+\delta_1} w \in H_{1+\delta_1}(0, T)$ for each $w \in L^2(0, T)$. Therefore

$$J^\alpha H_\delta(0, T) \subset H_{1+\delta_1}(0, T) = H_{\alpha+\delta}(0, T).$$

Thus (3.50) holds for all (α, β) satisfying $0 < \beta \leq 2$ and $0 < \alpha < 1$, and so the proof of (3.49) is complete. Thus the proof of Lemma 3.3 is finished.   ∎

3.6   Linear Fractional Ordinary Differential Equations with Multi-Term Fractional Derivatives

We have discussed the unique existence of the solutions to initial value problems for fractional ordinary differential equations on the basis of ∂_t^α in the Sobolev spaces $H_\alpha(0, T)$, and the method is widely applicable for example, to nonlinear equations. Rather than comprehensive discussions, here we are restricted to linear fractional ordinary differential equations with multi-term time fractional derivatives:

$$\sum_{j=1}^{m} r_j(t) \partial_t^{\alpha_j}(u - a)(t) = p(t) u(t) + f(t), \quad 0 < t < T \tag{3.51}$$

with

$$u - a \in H_{\alpha_1}(0, T). \tag{3.52}$$

Here

$$0 < \alpha_m < \alpha_{m-1} < \cdots < \alpha_1 < 1 \tag{3.53}$$

and

$$\begin{cases} r_1(t) > 0, \quad 0 \leq t \leq T, \quad r_1, \ldots, r_m \in L^\infty(0,T), \\ p \in L^\infty(0,T), \ f \in L^2(0,T). \end{cases} \tag{3.54}$$

Then we can prove:

Theorem 3.7 *We assume (3.53) and (3.54). There exists a unique solution u to (3.51) and (3.52), and with some constant $C > 0$, we have*

$$\|u\|_{L^2(0,T)} \leq C(|a| + \|f\|_{L^2(0,T)}), \tag{3.55}$$

$$\|u - a\|_{H^{\alpha_1}(0,T)} \leq C(|a| + \|F\|_{L^2(0,T)}) \tag{3.56}$$

and

$$\|u\|_{H^{\alpha_1}(0,T)} \leq C(|a| + \|f\|_{L^2(0,T)}), \quad if\, 0 < \alpha < 1\, and\, \alpha \neq \frac{1}{2}. \tag{3.57}$$

Proof We can rewrite (3.51) as

$$J^{-\alpha_1}(u - a)(t) + \sum_{j=2}^m \tilde{r}_j(t) J^{-\alpha_j}(u - a)(t) = \tilde{p}(t)u(t) + \tilde{f}(t), \quad 0 < t < T. \tag{3.58}$$

Here and henceforth we set

$$\tilde{r}_j(t) = \frac{r_j(t)}{r_1(t)}, \quad j = 2, \ldots, m, \quad \tilde{p}(t) = \frac{p(t)}{r_1(t)}, \quad \tilde{f}(t) = \frac{f(t)}{r_1(t)}, \quad 0 \leq t \leq T.$$

Therefore (3.51) with (3.52) is equivalent to

$$w(t) = -\sum_{j=2}^m \tilde{r}_j(t) J^{\alpha_1 - \alpha_j} w(t) + \tilde{p} J^{\alpha_1} w(t) + \tilde{p}(t)a + \tilde{f}(t), \quad 0 < t < T \tag{3.59}$$

and

$$w \in L^2(0,T). \tag{3.60}$$

The equivalence is directly seen by setting $w = J^{-\alpha_1}(u - a)$, because $u - a \in H^{\alpha_1}(0,T)$ if and only if $u - a = J^{\alpha_1}w$ with $w \in L^2(0,T)$ by Theorem 2.1.

Since $\alpha_1 - \alpha_j > 0$ for $j = 2, \ldots, m$, we see that $J^{\alpha_1 - \alpha_j} : L^2(0, T) \longrightarrow H_{\alpha_1 - \alpha_j}(0, T)$ by Theorem 2.2. By the compactness of the embedding:

$$H_{\alpha_1 - \alpha_j}(0, T) \longrightarrow L^2(0, T),$$

we verify that $J^{\alpha_1 - \alpha_j} : L^2(0, T) \longrightarrow L^2(0, T)$ is compact. Similarly $J^{\alpha_1} : L^2(0, T) \longrightarrow L^2(0, T)$ is compact, and by $\tilde{r}_j, \tilde{p} \in L^\infty(0, T)$, we see that the operator $-\sum_{j=2}^m \tilde{r}_j(t) J^{\alpha_1 - \alpha_j} + \tilde{p} J^{\alpha_1} : L^2(0, T) \longrightarrow L^2(0, T)$ is compact. Therefore if

$$v(t) = -\sum_{j=2}^m \tilde{r}_j(t) J^{\alpha_1 - \alpha_j} v(t) + \tilde{p} J^{\alpha_1} v(t), \quad 0 < t < T, \tag{3.61}$$

implies $v = 0$ in $(0, T)$, then the Fredholm alternative yields the unique existence of w satisfying (3.59) for any $a \in \mathbb{R}$ and $f \in L^2(0, T)$. Let (3.61) hold. Then

$$|v(t)| \le C \sum_{j=2}^m \int_0^t (t - s)^{\alpha_1 - \alpha_j - 1} |v(s)| ds + C \int_0^t (t - s)^{\alpha_1 - 1} |v(s)| ds$$

for $0 < t < T$. Since $\alpha_1 > \alpha_2 > \cdots > \alpha_m > 0$, we have

$$(t - s)^{\alpha_1 - \alpha_j - 1} = (t - s)^{\alpha_1 - \alpha_2 + (\alpha_2 - \alpha_j) - 1} \le T^{\alpha_2 - \alpha_j}(t - s)^{(\alpha_1 - \alpha_2) - 1}$$

and

$$(t - s)^{\alpha_1 - 1} = (t - s)^{(\alpha_1 - \alpha_2) - 1}(t - s)^{\alpha_2} \le T^{\alpha_2}(t - s)^{(\alpha_1 - \alpha_2) - 1},$$

and so

$$|v(t)| \le C \int_0^t (t - s)^{\alpha_1 - \alpha_2 - 1} |v(s)| ds, \quad 0 < t < T.$$

Noting that $\alpha_1 - \alpha_2 > 0$, we apply the generalized Gronwall inequality (Lemma A.2) to verify that $v = 0$ in $(0, T)$. Thus the proof of the unique existence is complete. The estimates (3.55)–(3.57) are proved similarly to Theorem 3.5. ∎

Chapter 4
Initial Boundary Value Problems for Time-Fractional Diffusion Equations

4.1 Main Results

Let $\Omega \subset \mathbb{R}^n$ be a bounded domain with boundary $\partial\Omega$ of C^2-class and let $\nu = \nu(x) = (\nu_1, \ldots, \nu_n)$ be the unit outward normal vector to $\partial\Omega$ at $x = (x_1, \ldots, x_n)$. Henceforth let

$$\partial_i = \frac{\partial}{\partial x_i}, \quad \partial_i^2 = \frac{\partial^2}{\partial x_i^2}, \quad i = 1, \ldots, n \quad \nabla = (\partial_1, \ldots, \partial_n), \quad \partial_s = \frac{\partial}{\partial s}.$$

We would like to discuss an initial boundary value problem for a time-fractional partial differential equation, which we can write as follows at the expense of rigor:

$$\begin{cases} d_t^\alpha u(x,t) + A(t)u(x,t) = F(x,t), & x \in \Omega, \ 0 < t \leq T, \\ u(x,t) = 0, & x \in \partial\Omega, \ 0 < t < T, \\ u(x,0) = a(x), & x \in \Omega. \end{cases} \tag{4.1}$$

Let $-A(t)$ be a uniform elliptic differential operator of the second order with (x,t)-dependent coefficients:

$$(-A(t)u)(x,t) = \sum_{i,j=1}^n \partial_i(a_{ij}(x,t)\partial_j u(x,t)) + \sum_{j=1}^n b_j(x,t)\partial_j u(x,t) + c(x,t)u(x,t),$$

for $x \in \Omega$ and $t > 0$, where $a_{ij} = a_{ji}, b_j, c \in C^2([0, T]; C^1(\overline{\Omega}))$, and there exists a constant $\mu_0 > 0$ such that

$$\sum_{i,j=1}^n a_{ij}(x, t)\xi_i\xi_j \geq \mu_0 \sum_{j=1}^n \xi_j^2, \quad x \in \overline{\Omega}, \, 0 \leq t < \infty, \, (\xi_1, \ldots, \xi_n) \in \mathbb{R}^n.$$

$$(4.2)$$

Remark 4.1 We can relax the regularity conditions of the coefficients similarly to [17, 30], by approximating them by smooth functions. However, we omit the details for simplicity.

The main purpose of this chapter is to formulate the initial boundary value problem for t-dependent $A(t)$ with initial value a and non-homogeneous term F in L^2-spaces. We refer to Kubica and Yamamoto [17], Zacher [30] for such treatments, and here we describe a unified approach within the framework by means of ∂_t^α in the Sobolev space $H_\alpha(0, T)$.

In particular, for more regular F, for example $F \in L^\infty(0, T; L^2(\Omega))$, there are several works on the well-posedness for fractional partial differential equations and we can refer to Gorenflo et al. [11], Li et al. [18], Luchko [21], Sakamoto and Yamamoto [26]. Here we do not intend any complete list of the references.

We mainly assume that a and F are in some L^2-spaces. Then in general we cannot prove that $u \in C([0, T]; L^2(\Omega))$. Therefore, similarly to the case of fractional ordinary differential equations, we must be careful in interpreting the initial condition $u(\cdot, 0) = a$ in (4.1), which cannot imply that $u(\cdot, t) \longrightarrow a$ in $L^2(\Omega)$ as $t \to 0$.

Henceforth for $H_0^1(\Omega) \subset L^2(\Omega)$, identifying the dual of $L^2(\Omega)'$ of $L^2(\Omega)$ with itself, we define $H^{-1}(\Omega) = (H_0^1(\Omega))'$. Then $H_0^1(\Omega) \subset L^2(\Omega) \subset H^{-1}(\Omega)$ algebraically and topologically (e.g., Brezis [5]).

In terms of the definition ∂_t^α in $H_\alpha(0, T)$ defined in Chap. 2, we formulate an initial boundary value problem as follows:

$$\partial_t^\alpha(u - a)(x, t) + A(t)u(x, t) = F \quad \text{in } H^{-1}(\Omega), \, 0 < t < T, \qquad (4.3)$$

$$u(\cdot, t) \in H_0^1(\Omega), \quad 0 < t < T, \qquad (4.4)$$

$$u - a \in H_\alpha(0, T; H^{-1}(\Omega)). \qquad (4.5)$$

For $\frac{1}{2} < \alpha < 1$, we have $H_\alpha(0, T) \subset H^\alpha(0, T) \subset C[0, T]$ by the Soloblev embedding, and so (4.5) implies that $u - a \in C([0, T]; H^{-1}(\Omega))$ and $u(x, 0) = a(x)$ is satisfied in the sense of $\lim_{t \to 0} u(\cdot, t) = a$ in $H^{-1}(\Omega)$.

Remark 4.2 Moreover, in view of (2.38), interpreting $\partial_t^\alpha u$ by (2.37), we note that (4.3) is equivalent to

$$\begin{cases} \partial_t^\alpha u + A(t)u = F + \frac{t^{-\alpha}}{\Gamma(1-\alpha)}a, & \text{in } H^{-1}(\Omega), \ 0 < t < T, \\ & \qquad\qquad \text{if } 0 < \alpha < \frac{1}{2}, \\ \partial_t^\alpha u + A(t)u = F, & \text{in } H^{-1}(\Omega), \ 0 < t < T, \\ & \qquad\qquad \text{if } \frac{1}{2} < \alpha < 1. \end{cases} \qquad (4.6)$$

We note that if $u \in C^1((0, T]; C(\overline{\Omega})) \cap C([0, T]; C(\overline{\Omega})) \cap L^2(0, T; C^2(\overline{\Omega}))$ satisfies (4.1), then u satisfies also (4.3)–(4.5).

Now we are ready to state our main results.

Theorem 4.1 *We assume regularity $a_{ij}, b_j, c \in C^2([0, T]; C^1(\overline{\Omega}))$, $1 \le i, j \le n$. For $F \in L^2(0, T; H^{-1}(\Omega))$ and $a \in L^2(\Omega)$, there exists a unique solution $u \in L^2(0, T; H_0^1(\Omega))$ satisfying $u - a \in H_\alpha(0, T; H^{-1}(\Omega))$ to (4.3)–(4.5). Moreover, there exists a constant $C > 0$ such that*

$$\|u - a\|_{H_\alpha(0,T;H^{-1}(\Omega))} + \|\nabla u\|_{L^2(0,T;L^2(\Omega))} \le C(\|a\|_{L^2(\Omega)} + \|F\|_{L^2(0,T;H^{-1}(\Omega))}). \qquad (4.7)$$

In Zacher [30], and Kubica and Yamamoto [17] (Theorem 1.1), with the same regularity conditions on a and F, the unique existence of the solution u to (4.3)–(4.5) is proved with the regularity

$$J^{1-\alpha}(u - a) \in {}_0H^1(0, T; H^{-1}(\Omega)), \quad u \in L^2(0, T; H_0^1(\Omega)).$$

Noting that ${}_0H^1(0, T; H^{-1}(\Omega)) = H_1(0, T; H^{-1}(\Omega))$ by the definition, we can verify that the class of solutions in [17, 30] is the same as in Theorem 4.1, that is,

$$J^{1-\alpha}(u - a) \in {}_0H^1(0, T; H^{-1}(\Omega)), \quad u \in L^2(0, T; H_0^1(\Omega))$$

if and only if

$$u - a \in H_\alpha(0, T; H^{-1}(\Omega)), \quad u \in L^2(0, T; H_0^1(\Omega)).$$

Indeed, setting $\alpha := 1 - \alpha$ and $\beta := \alpha$ in Theorem 2.3 (i) of Chap. 2, we see that $J^{1-\alpha} : H_\alpha(0, T) \longrightarrow H_1(0, T)$ is surjective and isomorphism. Therefore $J^{1-\alpha}(u - a) \in H_1(0, T; H^{-1}(\Omega)) = {}_0H^1(0, T; H^{-1}(\Omega))$ if and only if $u - u \in H_\alpha(0, T; H^{-1}(\Omega))$, which means that the solution classes coincide.

In particular, for $\frac{1}{2} < \alpha < 1$, we have

$$\|u - a\|_{C([0,T];H^{-1}(\Omega))} \le C\|F\|_{L^2(0,T;H^{-1}(\Omega))}. \qquad (4.8)$$

Indeed, the estimate (4.8) is directly seen by (4.7), because $H_\alpha(0, T) \subset H^\alpha(0, T) \subset C[0, T]$ if $\frac{1}{2} < \alpha < 1$ by the Sobolev embedding.

Remark 4.3 By (2.40), we can rewrite (4.7) as

$$\|\partial_t^\alpha u\|_{L^2(0,T;H^{-1}(\Omega))} + \|u\|_{L^2(0,T;H_0^1(\Omega))}$$

$$\leq C(\|a\|_{L^2(\Omega)} + \|F\|_{L^2(0,T;H^{-1}(\Omega))}), \quad \alpha \neq \frac{1}{2}. \tag{4.9}$$

Here ∂_t^α is defined by (2.37).

Proof of (4.9) For $0 < \alpha < \frac{1}{2}$, by (2.38) we have

$$\partial_t^\alpha (u - a) = \partial_t^\alpha u - \partial_t^\alpha a = \partial_t^\alpha u - \frac{t^{-\alpha}}{\Gamma(1 - \alpha)} a.$$

Therefore, since $\|t^{-\alpha}\|_{L^2(0,T)} < \infty$ for $0 < \alpha < \frac{1}{2}$, inequality (4.7) yields

$$\|\partial_t^\alpha u\|_{L^2(0,T;H^{-1}(\Omega))} \leq \|\partial_t^\alpha (u - a)\|_{L^2(0,T;H^{-1}(\Omega))} + \left\| \frac{t^{-\alpha}}{\Gamma(1 - \alpha)} a \right\|_{L^2(0,T;H^{-1}(\Omega))}$$

$$\leq C(\|F\|_{L^2(0,T;H^{-1}(\Omega))} + \|a\|_{L^2(0,T;H^{-1}(\Omega))}).$$

Next, let $\frac{1}{2} < \alpha < 1$. Then (4.9) is seen by the definition (2.37). ∎

Remark 4.4 In Sakamoto and Yamamoto [26], in a special case where $A(t)$ is symmetric (i.e., $b_j = c = 0, 1 \leq j \leq n$) and the coefficients of $A(t)$ are independent of t and $F = 0$, then for (4.1) it is proved that

$$\lim_{t \to 0} \|u(\cdot, t) - a\|_{L^2(\Omega)} = 0$$

for each $a \in L^2(\Omega)$. See also Sect. 4.5.

For more regular a and F, we can prove:

Proposition 4.1 *We assume all the conditions in Theorem 4.1. Moreover, let $p > \frac{2}{\alpha}$, and let $F \in L^p(0, T; H^{-1}(\Omega))$ and $a \in H_0^1(\Omega)$. Then there exists a constant $C > 0$ such that*

$$\|u - a\|_{L^\infty(0,T;L^2(\Omega))} \leq C(t^{\frac{\alpha}{2}} \|a\|_{H_0^1(\Omega)} + t^\kappa \|F\|_{L^p(0,T;H^{-1}(\Omega))}). \tag{4.10}$$

Here

$$\kappa = \frac{p\alpha - 2}{2p} > 0.$$

In particular,

$$\lim_{t \to 0} \|u(\cdot, t) - a\|_{L^2(\Omega)} = 0. \tag{4.11}$$

The proposition means that with more regular a and F, we can prove the continuity of $u(\cdot, t)$ at $t = 0$ in $L^2(\Omega)$, and so the initial condition (4.5) holds in the usual sense.

Remark 4.5 For $d_t^\alpha u = \text{div}(p(x, t)\nabla u(x, t)) + F(x, t)$ in (4.1) with $a = 0$, Jin, Li and Zhou (Theorem 2.1 in [12]) proves that $u \in C([0, T]; L^2(\Omega)) \cap L^p(0, T; H^2(\Omega))$ and $d_t^\alpha u \in L^p(0, T; L^2(\Omega))$ if $F \in L^p(0, T; L^2(\Omega))$ with $p > \frac{1}{\alpha}$. In fact, we can interpret that $p > \frac{1}{\alpha}$ is a critical condition for the continuity of u in t. Proposition 4.1 requires a stronger assumption $p > \frac{2}{\alpha}$ for p, but the weaker spatial regularity is needed. The continuity of u at $t = 0$ should be exploited more but here we omit details.

For more regular a and F, we can improve the regularity of u.

Theorem 4.2 *We assume regularity* $a_{ij} \in C^2([0, T]; C^1(\overline{\Omega}))$, $b_j, c \in C^2([0, T]; C^1(\overline{\Omega}))$, $1 \le i, j \le n$. *For* $F \in L^2(0, T; L^2(\Omega))$ *and* $a \in H_0^1(\Omega)$, *there exists a unique solution* $u \in L^2(0, T; H^2(\Omega) \cap H_0^1(\Omega))$ *satisfying* $u - a \in H_\alpha(0, T; L^2(\Omega))$ *to (4.3)–(4.5). Moreover, there exists a constant* $C > 0$ *such that*

$$\|u - a\|_{H_\alpha(0, T; L^2(\Omega))} + \|u\|_{L^2(0, T; H^2(\Omega))} \le C(\|F\|_{L^2(0, T; L^2(\Omega))} + \|a\|_{H_0^1(\Omega)}).$$

Theorems 4.1 and 4.2 are corresponding results to the classical results for the parabolic equation (i.e., $\alpha = 1$) for which we refer to Evans [7], Lions and Magenes [19], for example.

In the case of $F = 0$ in Theorem 4.2, we can further prove:

Proposition 4.2 *We assume all the conditions in Theorem 4.2 and* $F = 0$ *and*

$$\max_{1 \le i, j \le n} \|a_{ij}\|_{L^\infty(\Omega \times (0, \infty))}, \quad \max_{1 \le j \le n} \|b_j\|_{L^\infty(\Omega \times (0, \infty))}, \quad \|c\|_{L^\infty(\Omega \times (0, \infty))} < \infty.$$

Then there exists a constant $C > 0$ *such that*

$$\|u(\cdot, t)\|_{L^2(\Omega)} \le Ce^{Ct}\|u(\cdot, 0)\|_{H_0^1(\Omega)}, \quad t \ge 0$$

for all the solutions u to (4.3)–(4.5).

4.2 Some Results from Theorem 4.1 and Proposition 4.2

4.2.1 Interpolated Regularity of Solutions

Our classes of solutions in Theorems 4.1 and 4.2 are flexible for interpolated regularity properties. Although we can choose a general uniform elliptic operator, we introduce the Laplacian with the homogeneous Dirichlet boundary condition:

$$-A_0 u(x) = \sum_{k=1}^{n} \partial_k^2 u(x), \quad \mathcal{D}(A_0) = H^2(\Omega) \cap H_0^1(\Omega).$$

Then it is known that $\mathcal{D}(A_0^{\frac{1}{2}}) = H_0^1(\Omega)$ and $H^{-1}(\Omega) = (H_0^1(\Omega))' = \mathcal{D}(A_0^{-\frac{1}{2}})$. Moreover, we can verify that for $0 \le \theta \le 1$, the interpolation spaces are given by:

$$[H_\alpha(0, T; L^2(\Omega)), L^2(0, T; H^2(\Omega) \cap H_0^1(\Omega))]_{1-\theta}$$

$$= [H_\alpha(0, T; \mathcal{D}(A_0^0)), L^2(0, T; \mathcal{D}(A_0))]_{1-\theta}$$

$$= H_{\alpha\theta}(0, T; \mathcal{D}(A_0^{1-\theta})) \subset H_{\alpha\theta}(0, T; H^{2-2\theta}(\Omega))$$

(e.g., Yamamoto [29]). Here we consider only the case of $a = 0$ and $F \in L^2(0, T; L^2(\Omega))$.

Therefore for each $\theta \in [0, 1]$, in Theorem 4.2 we have $u \in H_{\alpha\theta}(0, T; H^{2-2\theta}(\Omega))$. In particular, by $\mathcal{D}(A_0^{\frac{1}{2}}) = H_0^1(\Omega)$, we see that $u \in H_{\frac{1}{2}\alpha}(0, T; H_0^1(\Omega))$ by choosing $\theta = \frac{1}{2}$ in Theorem 4.2, which is included in Theorem 1.4 in Kubica and Yamamoto [17].

4.2.2 The Method by the Laplace Transform for Fractional Partial Differential Equations

Let the elliptic operator $A(t)$ in (4.1) be t-independent:

$$(Av)(x) = -\sum_{i,j=1}^{n} \partial_i(a_{ij}(x)\partial_j v(x)) - \sum_{j=1}^{n} b_j(x)\partial_j v(x) - c(x)v(x), \quad x \in \Omega,$$

where $a_{ij}, b_j, c \in C^2(\overline{\Omega})$ and (4.2) is satisfied. A convenient way for constructing a solution to (4.3)–(4.5) is by the Laplace transform (see Sect. 2.7 of Chap. 2), which relies on formulae of Laplace transforms of time-fractional derivatives. For the rigorous treatments, we have to specify the class of solutions u admitting such formulae for the Laplace transforms, but it is not often clarified in view of the consistency with the expected regularity of solutions to (4.3)–(4.5).

Within our framework, we can apply Theorem 2.7 of Chap. 2 concerning the Laplace transform of ∂_t^α in $V_\alpha(0, \infty)$. Here we sketch such treatments. As related arguments, see Kian and Yamamoto [14].

For arbitrary $T > 0$, let $u \in L^2(0, T; H^2(\Omega) \cap H_0^1(\Omega))$ satisfy $u - a \in H_\alpha(0, T; L^2(\Omega))$ and (4.3)–(4.5) with $a \in H_0^1(\Omega)$ and $F = 0$. In particular,

$$\partial_t^\alpha(u - a) + Au = 0 \quad \text{in } \Omega \times (0, T).$$

Taking the scalar products in $L^2(\Omega)$ of both sides with arbitrarily fixed $\varphi \in C_0^\infty(\Omega)$ and integrating by parts, we have

$$(\partial_t^\alpha(u - a)(\cdot, t), \varphi)_{L^2(\Omega)} + (Au(\cdot, t), \varphi)_{L^2(\Omega)} = 0, \quad t > 0.$$

Then

$$\partial_t^\alpha\{((u - a)(\cdot, t), \varphi)_{L^2(\Omega)}\} + (u(\cdot, t), A^*\varphi)_{L^2(\Omega)} = 0, \quad t > 0. \tag{4.12}$$

Here

$$(A^*v)(x) = -\sum_{i,j=1}^n \partial_i(a_{ij}(x)\partial_j v) + \sum_{j=1}^n \partial_j(b_j(x)v(x)) - c(x)v(x), \quad x \in \Omega.$$

By $u - a \in H_\alpha(0, T; L^2(\Omega))$, we see that $((u - a)(\cdot, t), \varphi)_{L^2(\Omega)} \in H_\alpha(0, T)$. Proposition 4.2 yields $\|u(\cdot, t)\|_{L^2(\Omega)} \le Ce^{Ct}\|a\|_{H_0^1(\Omega)}$ for all $t \ge 0$. Hence

$$|((u - a)(\cdot, t), \varphi)_{L^2(\Omega)}|, \quad |(u(\cdot, t), A^*\varphi)_{L^2(\Omega)}| \le Ce^{C_1 t}, \quad t \ge 0.$$

Here $C > 0$ depends on a and φ.

Consequently, $((u(\cdot, t) - a), \varphi)_{L^2(\Omega)} \in V_\alpha(0, \infty)$ and the Laplace transform of $(u(\cdot, t), A^*\varphi)_{L^2(\Omega)}$:

$$L((u(\cdot, t), A^*\varphi)_{L^2(\Omega)})(p) := \int_0^\infty e^{-pt}(u(\cdot, t), A^*\varphi)_{L^2(\Omega)}dt$$

exists for $p > C_1$. We take the Laplace transforms of both sides of (4.12) and apply Theorem 2.7, so that

$$p^\alpha(L(u - a)(\cdot, p), \varphi)_{L^2(\Omega)} + ((Lu)(\cdot, p), A^*\varphi)_{L^2(\Omega)} = 0, \quad p > C_1,$$

that is,

$$(\{p^\alpha L(u - a)(\cdot, p) + A(Lu)(\cdot, p)\}, \varphi)_{L^2(\Omega)} = 0, \quad p > C_1$$

for all $\varphi \in C_0^\infty(\Omega)$, which means

$$p^\alpha L(u - a)(\cdot, p) + A(Lu)(\cdot, p) = 0, \quad p > C_1$$

in $L^2(\Omega)$. Thus we can justify the method by the Laplace transform for an initial boundary value problem (4.3)–(4.5).

4.3 Proof of Theorem 4.1

The proof is based on what is called the Galerkin approximation (e.g., Evans [7], Lions and Magenes [19]). The proof is composed of:

(i) Construction of approximating solutions in a family of finitely dimensional subspaces: For this step, Theorem 3.6 (iii) plays an important role.
(ii) Uniform boundedness of the approximating solutions: The coercivity Theorems 3.3 and 3.4 are essential.
(iii) Then we can prove the existence of solution as a weak convergent limit of a subsequence of the sequence of the approximating solutions. The uniqueness of the solution follows from the coercivity.

First we prove the existence of solutions.

Let $0 < \lambda_1 \leq \lambda_2 \leq \cdots$ be the eigenvalues of $-\Delta$ with the zero Dirichlet boundary condition, which are numbered according to the multiplicities, that is, λ_m appears ℓ-times if the multiplicity of λ_m is ℓ. By $\rho_k \in H^2(\Omega), k \in \mathbb{N}$, we denote an eigenfunction of $-\Delta$ for λ_k such that

$$(\rho_k, \rho_\ell)_{L^2(\Omega)} = \delta_{k\ell} := \begin{cases} 1, & k = \ell, \\ 0, & k \neq \ell. \end{cases}$$

Then

$$\{\rho_k\}_{k\in\mathbb{N}} \text{ is a basis in each } L^2(\Omega), H_0^1(\Omega),$$

$$(\rho_k, \rho_\ell)_{L^2(\Omega)} = \delta_{k\ell}, \quad {}_{H^{-1}(\Omega)}\langle \rho_k, \rho_\ell \rangle_{H_0^1(\Omega)} = \delta_{k\ell}. \tag{4.13}$$

Moreover,

$$\begin{cases} a = \sum_{k=1}^\infty (a, \rho_k)_{L^2(\Omega)} \rho_k \text{ in } L^2(\Omega), \quad \|a\|_{L^2(\Omega)}^2 = \sum_{k=1}^\infty |(a, \rho_k)_{L^2(\Omega)}|^2, \\ \text{there exists a constant } C > 0 \text{ such that} \\ a = \sum_{k=1}^\infty {}_{H^{-1}(\Omega)}\langle a, \rho_k \rangle_{H_0^1(\Omega)} \rho_k \text{ in } H^{-1}(\Omega) \text{ and} \\ C^{-1} \sum_{k=1}^\infty \lambda_k^{-1} |{}_{H^{-1}(\Omega)}\langle a, \rho_k \rangle_{H_0^1(\Omega)}|^2 \leq \|a\|_{H^{-1}(\Omega)}^2 \\ \leq C \sum_{k=1}^\infty \lambda_k^{-1} |{}_{H^{-1}(\Omega)}\langle a, \rho_k \rangle_{H_0^1(\Omega)}|^2, \quad a \in H^{-1}(\Omega), \\ C^{-1} \sum_{k=1}^\infty \lambda_k |(a, \rho_k)|^2 \leq \|a\|_{H_0^1(\Omega)}^2 \leq C \sum_{k=1}^\infty \lambda_k |(a, \rho_k)|^2, \quad a \in H_0^1(\Omega). \end{cases}$$
$$\tag{4.14}$$

We note that

$$_{H^{-1}(\Omega)}\langle \varphi, \psi \rangle_{H_0^1(\Omega)} = (\varphi, \psi)_{L^2(\Omega)} \quad \text{for } \varphi \in L^2(\Omega) \text{ and } \psi \in H_0^1(\Omega)$$

(e.g., p.136 in [5]).

By the density of $C_0^\infty(0, T; H^{-1}(\Omega))$ in $L^2(0, T; H^{-1}(\Omega))$, for any $\varepsilon > 0$, we can choose $F^\varepsilon \in C_0^\infty(0, T; H^{-1}(\Omega))$ such that

$$\|F^\varepsilon - F\|_{L^2(0,T;H^{-1}(\Omega))} < \varepsilon. \tag{4.15}$$

Indeed, we can construct such F^ε by the mollifier $\int_0^T F(x, s)\chi_\varepsilon(t - s)ds$, where $\chi_\varepsilon \in C_0^\infty(\mathbb{R}), \geq 0$ satisfies supp $\chi_\varepsilon \subset (-\varepsilon, \varepsilon)$ and $\int_{-\infty}^\infty \chi_\varepsilon(t)dt = 1$ for small $\varepsilon > 0$.

Let $N \in \mathbb{N}$ be fixed arbitrarily. We look for

$$u_N^\varepsilon(x, t) = \sum_{k=1}^N p_{N,k}^\varepsilon(t)\rho_k(x), \quad F_N^\varepsilon(x, t) = \sum_{k=1}^N {}_{H^{-1}(\Omega)}\langle F^\varepsilon(\cdot, t), \rho_k \rangle_{H_0^1(\Omega)} \rho_k(x) \tag{4.16}$$

satisfying

$$\begin{cases} (\partial_t^\alpha(u_N^\varepsilon - a_N), \rho_\ell)_{L^2(\Omega)} + (A(t)u_N^\varepsilon, \rho_\ell)_{L^2(\Omega)} = {}_{H^{-1}(\Omega)}\langle F_N^\varepsilon, \rho_\ell \rangle_{H_0^1(\Omega)}, \\ u_N^\varepsilon - a_N \in H_\alpha(0, T; L^2(\Omega)), \quad 1 \leq \ell \leq N, 0 \leq t \leq T, \end{cases} \tag{4.17}$$

where we set

$$a_N = \sum_{k=1}^N c_k\rho_k, \quad c_k = (a, \rho_k)_{L^2(\Omega)}. \tag{4.18}$$

Multiplying the first equation in (4.17) by $p_{N,\ell}^\varepsilon$ and summing up over $\ell = 1, \ldots, N$, we have

$$(\partial_t^\alpha(u_N^\varepsilon - a_N), u_N^\varepsilon)_{L^2(\Omega)} + (A(t)u_N^\varepsilon, u_N^\varepsilon)_{L^2(\Omega)}$$

$$= {}_{H^{-1}(\Omega)}\langle F_N^\varepsilon, u_N^\varepsilon \rangle_{H_0^1(\Omega)}, \quad 0 \leq t \leq T. \tag{4.19}$$

We rewrite (4.17) as

$$\begin{cases} \partial_t^\alpha(p_{N,\ell}^\varepsilon - c_\ell) = \sum_{k=1}^N \Lambda_{k\ell}(t)p_{N,k}^\varepsilon(t) + f_\ell^\varepsilon(t), \quad 0 < t < T, \\ p_{N,\ell}^\varepsilon - c_\ell \in H_\alpha(0, T), \quad 1 \leq \ell \leq N, \end{cases} \tag{4.20}$$

where

$$\begin{cases} f_\ell^\varepsilon(t) = {}_{H^{-1}(\Omega)}\langle F^\varepsilon(\cdot,t), \rho_\ell\rangle_{H_0^1(\Omega)}, \\ \Lambda_{k\ell}(t) = \sum_{i,j=1}^n (\partial_i(a_{ij}(\cdot,t)\partial_j\rho_k), \rho_\ell(\cdot))_{L^2(\Omega)} \\ + \sum_{i=1}^n (b_i(\cdot,t)\partial_i\rho_k, \rho_\ell)_{L^2(\Omega)} + (c(\cdot,t)\rho_k, \rho_\ell)_{L^2(\Omega)} \\ = -\sum_{i,j=1}^n \int_\Omega a_{ij}(x,t)(\partial_j\rho_k)(x)(\partial_i\rho_\ell)(x)dx \\ + \sum_{i=1}^n (b_i(\cdot,t)\partial_i\rho_k, \rho_\ell)_{L^2(\Omega)} + (c(\cdot,t)\rho_k, \rho_\ell)_{L^2(\Omega)}. \end{cases}$$

By the regularity of F^ε and a_{ij}, we know that $\Lambda_{k\ell} \in C^2[0, T]$ and $f_\ell^\varepsilon \in W^{1,1}(0, T)$. Consequently, we can apply Theorem 3.6 (iii) to see that there exists a unique solution $p_{N,\ell}^\varepsilon \in W^{1,1}(0, T)$ to (4.20) and $p_{N,\ell}^\varepsilon(0) = c_\ell$ for $\ell = 1, \ldots, N$.

Next we estimate u_N^ε Since $p_{N,k}^\varepsilon \in W^{1,1}(0, T)$ and $p_{N,k}^\varepsilon(0) = c_k$, we see that $p_{N,k}^\varepsilon - c_k \in H_\alpha(0, T) \cap {}_0W^{1,1}(0, T)$ and so (2.27) yields

$$\partial_t^\alpha(p_{N,k}^\varepsilon - c_k) = d_t^\alpha(p_{N,k}^\varepsilon - c_k) = d_t^\alpha p_{N,k}^\varepsilon, \quad N \in \mathbb{N}, 1 \le k \le N. \tag{4.21}$$

Therefore,

$$\partial_t^\alpha(u_N^\varepsilon - a_N) = \sum_{k=1}^N \partial_t^\alpha(p_{N,k}^\varepsilon - c_k)\rho_k = \sum_{k=1}^N (d_t^\alpha p_{N,k}^\varepsilon)\rho_k = d_t^\alpha u_N^\varepsilon. \tag{4.22}$$

By (4.19) we have

$$(t-s)^{\alpha-1}(d_t^\alpha u_N^\varepsilon(\cdot,s), u_N^\varepsilon(\cdot,s))_{L^2(\Omega)}$$

$$- \sum_{i,j=1}^n (t-s)^{\alpha-1}(\partial_i(a_{ij}(\cdot,s)\partial_j u_N^\varepsilon(\cdot,s)), u_N^\varepsilon(\cdot,s))_{L^2(\Omega)}$$

$$- \sum_{j=1}^n (t-s)^{\alpha-1}(b_j(\cdot,s)\partial_j u_N^\varepsilon(\cdot,s), u_N^\varepsilon(\cdot,s))_{L^2(\Omega)}$$

$$- (t-s)^{\alpha-1}(c(\cdot,s)u_N^\varepsilon(\cdot,s), u_N^\varepsilon(\cdot,s))_{L^2(\Omega)}$$

$$= {}_{H^{-1}(\Omega)}\langle F_N^\varepsilon(\cdot,s), u_N^\varepsilon(\cdot,s)\rangle_{H_0^1(\Omega)}(t-s)^{\alpha-1}. \tag{4.23}$$

By (4.2) we obtain

$$- \sum_{i,j=1}^n (t-s)^{\alpha-1}(\partial_i(a_{ij}(\cdot,s)\partial_j u_N^\varepsilon(\cdot,s)), u_N^\varepsilon(\cdot,s))_{L^2(\Omega)}$$

$$= \sum_{i,j=1}^n (t-s)^{\alpha-1} \int_\Omega a_{ij}(x,s)(\partial_j u_N^\varepsilon(x,s))\partial_i u_N^\varepsilon(x,s))dx$$

$$\ge C_0(t-s)^{\alpha-1}\|\nabla u_N^\varepsilon(\cdot,s)\|_{L^2(\Omega)}^2. \tag{4.24}$$

Moreover, fixing small $\delta > 0$, we can choose a constant $C_\delta > 0$ such that we have

$$\left| -\sum_{j=1}^{n} (b_j(\cdot, s)\partial_j u_N^\varepsilon(\cdot, s), u_N^\varepsilon(\cdot, s))_{L^2(\Omega)} \right| \leq C \int_\Omega \sum_{j=1}^{n} |\partial_j u_N^\varepsilon(x, s)||u_N^\varepsilon(x, s)|dx$$

$$\leq \delta \int_\Omega |\nabla u_N^\varepsilon(x, s)|^2 dx + C_\delta \int_\Omega |u_N^\varepsilon(x, s)|^2 dx. \tag{4.25}$$

Similarly we see

$$|_{H^{-1}(\Omega)}\langle F_N^\varepsilon(\cdot, s), u_N^\varepsilon(\cdot, s)\rangle_{H_0^1(\Omega)}| \leq \delta\|\nabla u_N^\varepsilon(\cdot, s)\|_{L^2(\Omega)}^2 + C_\delta\|F_N^\varepsilon(\cdot, s)\|_{H^{-1}(\Omega)}^2. \tag{4.26}$$

On the other hand, by Theorem 3.4 (i) and $u_N^\varepsilon \in W^{1,1}(0, T; L^2(\Omega))$, we obtain

$$\int_0^t (t-s)^{\alpha-1}(d_t^\alpha u_N^\varepsilon(\cdot, s), u_N^\varepsilon(\cdot, s))_{L^2(\Omega)}ds \geq \frac{\Gamma(\alpha)}{2}(\|u_N^\varepsilon(\cdot, t)\|_{L^2(\Omega)}^2 - \|a_N\|_{(L^2(\Omega))^N}^2).$$

Therefore application of (4.24)–(4.26) in (4.23) yields

$$\|u_N^\varepsilon(\cdot, t)\|_{L^2(\Omega)}^2 + \int_0^t (t-s)^{\alpha-1}\|\nabla u_N^\varepsilon(\cdot, s)\|_{L^2(\Omega)}^2 ds$$

$$\leq \|a_N\|_{L^2(\Omega)}^2 + C\delta \int_0^t (t-s)^{\alpha-1}\|\nabla u_N^\varepsilon(\cdot, s)\|_{L^2(\Omega)}^2 ds$$

$$+ C_\delta \int_0^t (t-s)^{\alpha-1}\|u_N^\varepsilon(\cdot, s)\|_{L^2(\Omega)}^2 ds + C_\delta \int_0^t (t-s)^{\alpha-1}\|F_N^\varepsilon(\cdot, s)\|_{H^{-1}(\Omega)}^2 ds, \quad 0 < t < T.$$

Fixing $\delta > 0$ sufficiently small and absorbing the second term on the right-had side into the left-hand side, we obtain

$$\|u_N^\varepsilon(\cdot, t)\|_{L^2(\Omega)}^2$$

$$\leq \|u_N^\varepsilon(\cdot, t)\|_{L^2(\Omega)}^2 + \int_0^t (t-s)^{\alpha-1}\|\nabla u_N^\varepsilon(\cdot, s)\|_{L^2(\Omega)}^2 ds$$

$$\leq C\|a_N\|_{L^2(\Omega)}^2 + C \int_0^t (t-s)^{\alpha-1}\|u_N^\varepsilon(\cdot, s)\|_{L^2(\Omega)}^2 ds$$

$$+ C \int_0^t (t-s)^{\alpha-1}\|F_N^\varepsilon(\cdot, s)\|_{H^{-1}(\Omega)}^2 ds. \tag{4.27}$$

The generalized Gronwall inequality Lemma A.2 in the Appendix, implies

$$\|u_N^\varepsilon(\cdot, t)\|_{L^2(\Omega)}^2$$

$$\leq C\|a_N\|_{L^2(\Omega)}^2 + C \int_0^t (t-s)^{\alpha-1}\|F_N^\varepsilon(\cdot, s)\|_{H^{-1}(\Omega)}^2 ds$$

$$+ C \int_0^t (t-s)^{\alpha-1}\left(\int_0^s (s-\xi)^{\alpha-1}\|F_N^\varepsilon(\cdot, \xi)\|_{H^{-1}(\Omega)}^2 d\xi\right) ds, \quad 0 < t < T.$$

Here

$$\int_0^t (t-s)^{\alpha-1}\left(\int_0^s (s-\xi)^{\alpha-1}\|F_N^\varepsilon(\cdot, \xi)\|_{H^{-1}(\Omega)}^2 d\xi\right) ds$$

$$= \int_0^t \|F_N^\varepsilon(\cdot, \xi)\|_{H^{-1}(\Omega)}^2 \left(\int_\xi^t (t-s)^{\alpha-1}(s-\xi)^{\alpha-1} ds\right) d\xi$$

$$= \frac{\Gamma(\alpha)^2}{\Gamma(2\alpha)} \int_0^t (t-\xi)^{2\alpha-1}\|F_N^\varepsilon(\cdot, \xi)\|_{H^{-1}(\Omega)}^2 d\xi.$$

Since

$$(t-s)^{2\alpha-1} \leq C(t-s)^{\alpha-1}, \quad 0 \leq s \leq t \leq T,$$

we have

$$\|u_N^\varepsilon(\cdot, t)\|_{L^2(\Omega)}^2 \leq C\|a_N\|_{L^2(\Omega)}^2 + C \int_0^t (t-s)^{\alpha-1}\|F_N^\varepsilon(\cdot, s)\|_{H^{-1}(\Omega)}^2 ds.$$

Taking $\int_0^T \cdots dt$ and applying the Young inequality on the convolution (Lemma A.1 in the Appendix), we have

$$\|u_N^\varepsilon(\cdot, t)\|_{L^2(0,T;L^2(\Omega))}^2 \leq C(\|a_N\|_{L^2(\Omega)}^2 + \|F_N^\varepsilon\|_{L^2(0,T;H^{-1}(\Omega))}^2). \tag{4.28}$$

By (4.22), applying Theorem 3.3 (i) we obtain

$$\int_0^T (\partial_t^\alpha(u_N^\varepsilon - a_N)(\cdot, t), u_N^\varepsilon(\cdot, t))_{L^2(\Omega)} dt = \int_0^T (d_t^\alpha u_N^\varepsilon(\cdot, t), u_N^\varepsilon(\cdot, t))_{L^2(\Omega)} dt$$

$$\geq C_0 \left(\int_0^T \|u_N^\varepsilon(\cdot, t)\|_{L^2(\Omega)}^2 dt - \|a_N\|_{L^2(\Omega)}^2\right). \tag{4.29}$$

On the other hand, estimating the second term on the left-hand side and the right-hand side of (4.19) by (4.29) and (4.25)–(4.26), we have

$$\int_0^T \|u_N^\varepsilon(\cdot,t)\|_{L^2(\Omega)}^2 dt + \int_0^T \|\nabla u_N^\varepsilon(\cdot,t)\|_{L^2(\Omega)}^2 dt$$

$$\leq C\delta \int_0^T \|\nabla u_N^\varepsilon(\cdot,t)\|_{L^2(\Omega)}^2 dt + C_\delta \int_0^T \|u_N^\varepsilon(\cdot,t)\|_{L^2(\Omega)}^2 dt$$

$$+ C_\delta \int_0^T \|F_N^\varepsilon(\cdot,t)\|_{H^{-1}(\Omega)}^2 dt + C\|a_N\|_{L^2(\Omega)}^2.$$

Therefore, fixing $\delta > 0$ small and absorbing the first term on the right-hand side into the left-hand side, we reach

$$\|\nabla u_N^\varepsilon\|_{L^2(0,T;L^2(\Omega))}^2 \leq C(\|a_N\|_{L^2(\Omega)}^2 + \|u_N^\varepsilon\|_{L^2(0,T;L^2(\Omega))}^2 + \|F_N^\varepsilon\|_{L^2(0,T;H^{-1}(\Omega))}^2).$$

Applying (4.28), we obtain

$$\|\nabla u_N^\varepsilon\|_{L^2(0,T;L^2(\Omega))}^2 \leq C(\|a_N\|_{L^2(\Omega)}^2 + \|F_N^\varepsilon\|_{L^2(0,T;H^{-1}(\Omega))}^2). \tag{4.30}$$

Here $C > 0$ is independent of $N \in \mathbb{N}$ and $\varepsilon > 0$.

Now we estimate $\|\partial_t^\alpha(u_N^\varepsilon - a_N)\|_{L^2(0,T;H^{-1}(\Omega))}$. Since

$$\partial_t^\alpha(u_N^\varepsilon - a_N) = \sum_{k=1}^N \partial_t^\alpha(p_{N,k}^\varepsilon - c_k)\rho_k,$$

we have

$$_{H^{-1}(\Omega)}\langle \partial_t^\alpha(u_N^\varepsilon - a_N), \rho_\ell \rangle_{H_0^1(\Omega)} = \sum_{k=1}^N \partial_t^\alpha(p_{N,k}^\varepsilon(t) - c_k)_{H^{-1}(\Omega)}\langle \rho_k, \rho_\ell \rangle_{H_0^1(\Omega)} = 0$$

for $\ell \geq N + 1$. For any $\psi \in H_0^1(\Omega)$, we set $\psi_N = \sum_{\ell=1}^N {}_{H^{-1}(\Omega)}\langle \psi, \rho_\ell \rangle_{H_0^1(\Omega)}\rho_\ell$. Therefore (4.17) yields

$$_{H^{-1}(\Omega)}\langle \partial_t^\alpha(u_N^\varepsilon - a_N), \psi \rangle_{H_0^1(\Omega)}$$

$$= -{}_{H^{-1}(\Omega)}\langle A(t)u_N^\varepsilon, \psi_N \rangle_{H_0^1(\Omega)} + {}_{H^{-1}(\Omega)}\langle F_N^\varepsilon, \psi_N \rangle_{H_0^1(\Omega)}, \quad \psi \subset H_0^1(\Omega).$$

Hence

$$\|\partial_t^\alpha (u_N^\varepsilon - a_N)(\cdot, t)\|_{H^{-1}(\Omega)} = \sup_{\psi \in H_0^1(\Omega), \|\psi\|_{H_0^1(\Omega)}=1} |_{H^{-1}(\Omega)} \langle \partial_t^\alpha (u_N^\varepsilon - a_N)(\cdot, t), \psi \rangle_{H_0^1(\Omega)}|$$

$$\leq \sup_{\psi \in H_0^1(\Omega), \|\psi\|_{H_0^1(\Omega)}=1} |_{H^{-1}(\Omega)} \langle A(t)u_N^\varepsilon, \psi_N \rangle_{H_0^1(\Omega)}|$$

$$+ \sup_{\psi \in H_0^1(\Omega), \|\psi\|_{H_0^1(\Omega)}=1} |_{H^{-1}(\Omega)} \langle F_N^\varepsilon(\cdot, t), \psi_N \rangle_{H_0^1(\Omega)}|$$

$$\leq C \sup_{\psi \in H_0^1(\Omega), \|\psi\|_{H_0^1(\Omega)}=1} |(\nabla u_N^\varepsilon(\cdot, t), \nabla \psi_N)_{L^2(\Omega)}|$$

$$+ \sup_{\psi \in H_0^1(\Omega), \|\psi\|_{H_0^1(\Omega)}=1} \|F_N^\varepsilon(\cdot, t)\|_{H^{-1}(\Omega)} \|\nabla \psi_N\|_{L^2(\Omega)}$$

$$\leq C(\|u_N^\varepsilon(\cdot, t)\|_{L^2(\Omega)} + \|F_N^\varepsilon(\cdot, t)\|_{H^{-1}(\Omega)}).$$

Here we used also $\|\psi_N\|_{H_0^1(\Omega)} \leq \|\psi\|_{H_0^1(\Omega)} \leq 1$.

Since $\|a_N\|_{L^2(\Omega)} \leq \|a\|_{L^2(\Omega)}$ and $\|F_N^\varepsilon\|_{L^2(0,T;H^{-1}(\Omega))} \leq \|F^\varepsilon\|_{L^2(0,T;H^{-1}(\Omega))}$, combining (4.28) and (4.30), we obtain

$$\|\partial_t^\alpha (u_N^\varepsilon - a_N)\|_{L^2(0,T;H^{-1}(\Omega))}^2 + \|u_N^\varepsilon\|_{L^2(0,T;H_0^1(\Omega))}^2$$

$$\leq C(\|a\|_{L^2(\Omega)}^2 + \|F^\varepsilon\|_{L^2(0,T;H^{-1}(\Omega))}^2). \tag{4.31}$$

The sequences $\{u_N^\varepsilon - a_N\}_{N \in \mathbb{N}}$ and $\{u_N^\varepsilon\}_{N \in \mathbb{N}}$ are bounded in $H^\alpha(0, T; H^{-1}(\Omega))$ and in $L^2(0, T; H_0^1(\Omega))$ respectively. Therefore we can extract a subsequence N' of $N \in \mathbb{N}$ and $u^\varepsilon \in L^2(0, T; H_0^1(\Omega))$ and $v^\varepsilon \in H_\alpha(0, T; H^{-1}(\Omega))$ such that $u_{N'}^\varepsilon \longrightarrow u^\varepsilon$ weakly in $L^2(0, T; H_0^1(\Omega))$ and $u_{N'}^\varepsilon - a_{N'} \longrightarrow v^\varepsilon$ weakly in $H_\alpha(0, T; H^{-1}(\Omega))$. Since $a_{N'} \longrightarrow a$ strongly in $L^2(\Omega)$, we have $u_{N'}^\varepsilon \longrightarrow a + v^\varepsilon$ weakly in $H_\alpha(0, T; H^{-1}(\Omega))$. Therefore $u_{N'}^\varepsilon \longrightarrow u^\varepsilon$ and $u_{N'}^\varepsilon \longrightarrow a + v^\varepsilon$ in the sense of distribution, that is, in $(C_0^\infty(\Omega \times (0, T)))'$. Hence $u^\varepsilon = a + v^\varepsilon$, that is, $v^\varepsilon = u^\varepsilon - a$. Therefore by (4.31) we have

$$\|u^\varepsilon - a\|_{H_\alpha(0,T;H^{-1}(\Omega))}^2 + \|u^\varepsilon\|_{L^2(0,T;H_0^1(\Omega))}^2 \tag{4.32}$$

$$\leq \liminf_{N' \to \infty} \left(\|u_{N'}^\varepsilon - a_{N'}\|_{H^\alpha(0,T;H^{-1}(\Omega))}^2 + \|u_{N'}^\varepsilon\|_{L^2(0,T;H_0^1(\Omega))}^2 \right)$$

$$\leq C(\|F^\varepsilon\|_{L^2(0,T;H^{-1}(\Omega))}^2 + \|a\|_{L^2(\Omega)}^2).$$

Henceforth by X_N we denote the closed subspace in $H_0^1(\Omega)$ spanned by $\rho_1, \ldots, \rho_N$. For any $\psi \in X_N$, it follows from (4.17) that

$$\int_0^T {}_{H^{-1}(\Omega)}\langle \partial_t^\alpha (u_N^\varepsilon - a_N), \psi \rangle_{H_0^1(\Omega)} dt + \int_0^T {}_{H^{-1}(\Omega)}\langle A(t)u_N^\varepsilon, \psi \rangle_{H_0^1(\Omega)} dt$$

$$= \int_0^T {}_{H^{-1}(\Omega)}\langle F_N^\varepsilon(\cdot, t), \psi \rangle_{H_0^1(\Omega)} dt, \quad N \in \mathbb{N}.$$

Consequently, letting $N \to \infty$, we obtain

$$\int_0^T {}_{H^{-1}(\Omega)}\langle \partial_t^\alpha (u^\varepsilon - a), \psi \rangle_{H_0^1(\Omega)} dt + \int_0^T {}_{H^{-1}(\Omega)}\langle A(t)u, \psi \rangle_{H_0^1(\Omega)} dt$$

$$= \int_0^T {}_{H^{-1}(\Omega)}\langle F^\varepsilon(\cdot, t), \psi \rangle_{H_0^1(\Omega)} dt. \tag{4.33}$$

Since $\psi_N \in X_N$ and $N \in \mathbb{N}$ are chosen arbitrarily, it follows that (4.33) holds for each $\psi \in H_0^1(\Omega)$, so that

$$\partial_t^\alpha (u^\varepsilon - a) + A(t)u^\varepsilon = F^\varepsilon \quad \text{in } H^{-1}(\Omega), 0 < t < T.$$

By (4.32) we choose a sequence $\{\varepsilon_n\}_{n\in\mathbb{N}}$ satisfying $\lim_{n\to\infty} \varepsilon_n = 0$ and $u \in L^2(0, T; H_0^1(\Omega))$ such that $\partial_t^\alpha (u^{\varepsilon_n} - a) \longrightarrow \partial_t^\alpha (u - a)$ weakly in $L^2(0, T; H^{-1}(\Omega))$ and $u^{\varepsilon_n} \longrightarrow u$ weakly in $L^2(0, T; H_0^1(\Omega))$ as $n \to \infty$. Since $F_{\varepsilon_n} \longrightarrow F$ in $L^2(0, T; H^{-1}(\Omega))$ as $n \to \infty$ by (4.15), we can verify (4.7) and

$$\partial_t^\alpha (u - a)(\cdot, t) + A(t)u(\cdot, t) = F(\cdot, t) \quad \text{in } H^{-1}(\Omega), 0 < t < T.$$

Finally, we have to prove the uniqueness of the solution to (4.3)–(4.5). To this end, it is necessary to verify that if $u \in H_\alpha(0, T; H^{-1}(\Omega)) \cap L^2(0, T; H_0^1(\Omega))$ satisfies

$$\partial_t^\alpha u - \sum_{i,j=1}^n \partial_i (a_{ij}(x, t)\partial_j u) = \sum_{j=1}^n b_j(x, t)\partial_j u + c(x, t)u \quad \text{in } H^{-1}(\Omega), \tag{4.34}$$

then $u = 0$ in $\Omega \times (0, T)$.

From (4.34) and integration by parts in view of $u(\cdot, t) \in H_0^1(\Omega)$, we have

$${}_{H^{-1}(\Omega)}\langle \partial_t^\alpha u(\cdot, s), u(\cdot, s) \rangle_{H_0^1(\Omega)} + \left(\sum_{i,j=1}^n a_{ij}(\cdot, s)\partial_j u(\cdot, s), \partial_i u(\cdot, s) \right)_{L^2(\Omega)}$$

$$= {}_{H^{-1}(\Omega)}\langle \sum_{j=1}^n b_j(\cdot, s)\partial_j u(\cdot, s), u(\cdot, s) \rangle_{H_0^1(\Omega)} + {}_{H^{-1}(\Omega)}\langle c(\cdot, s)u, u \rangle_{H_0^1(\Omega)}, \quad 0 < s < T.$$

Therefore

$$\int_0^t {}_{H^{-1}(\Omega)}\langle \partial_t^\alpha u(\cdot, s), u(\cdot, s)\rangle_{H_0^1(\Omega)}(t-s)^{\alpha-1}ds$$

$$+ \int_0^t (t-s)^{\alpha-1}\left(\sum_{i,j=1}^n a_{ij}(x,s)(\partial_j u(x,s))\partial_i u(x,s)dx\right)ds$$

$$= \int_0^t (t-s)^{\alpha-1}\left(\int_\Omega \sum_{j=1}^n b_j(x,s)(\partial_j u(x,s))u(x,s)dx\right)ds$$

$$+ \int_0^t (t-s)^{\alpha-1}\left(\int_\Omega c(x,s)u^2(x,s)dx\right)ds.$$

Applying (4.2) and (4.25), we derive

$$\int_0^t {}_{H^{-1}(\Omega)}\langle \partial_t^\alpha u(\cdot, s), u(\cdot, s)\rangle_{H_0^1(\Omega)}(t-s)^{\alpha-1}ds$$

$$+ \int_0^t (t-s)^{\alpha-1}\|\nabla u(\cdot, s)\|_{L^2(\Omega)}^2 ds$$

$$\leq C\delta \int_0^t (t-s)^{\alpha-1}\|\nabla u(\cdot, s)\|_{L^2(\Omega)}^2 ds + C_\delta \int_0^t (t-s)^{\alpha-1}\|u(\cdot, s)\|_{L^2(\Omega)}^2 ds.$$

By $u \in H_\alpha(0, T; H^{-1}(\Omega)) \cap L^2(0, T; H_0^1(\Omega))$, applying Theorem 3.4 (ii) to the first term on the left-hand side and choosing $\delta > 0$ sufficiently small, we have

$$\frac{\Gamma(\alpha)}{2}\|u(\cdot, t)\|_{L^2(\Omega)}^2 \leq \frac{\Gamma(\alpha)}{2}\|u(\cdot, t)\|_{L^2(\Omega)}^2 + (1 - C\delta)\int_0^t (t-s)^{\alpha-1}\|\nabla u(\cdot, s)\|_{L^2(\Omega)}^2 ds$$

$$\leq C_\delta \int_0^t (t-s)^{\alpha-1}\|u(\cdot, s)\|_{L^2(\Omega)}^2 ds, \quad 0 < t < T.$$

The generalized Gronwall inequality Lemma A.2 in the Appendix implies $u = 0$ in $(0, T)$. Thus the proof of Theorem 4.1 is complete.

4.4   Proof of Theorem 4.2

For $F \in L^2(0, T; L^2(\Omega))$ and $\varepsilon > 0$, let $F^\varepsilon \in C_0^\infty(0, T; L^2(\Omega))$ satisfy $\|F^\varepsilon - F\|_{L^2(0,T;L^2(\Omega))} < \varepsilon$. In terms of Theorem 3.6 (iii) and (v), we can argue similarly to the proof of Theorem 4.1 to construct an approximating sequence, that is, $p_{N,k}^\varepsilon \in$

$(W_\alpha(0, T))^N$ and $p^\varepsilon_{N,k}(0) = c_k := (a, \rho_k)_{L^2(\Omega)}$ for $1 \le k \le N$. Here we have to estimate

$$u^\varepsilon_N(x, t) := \sum_{k=1}^N p^\varepsilon_{N,k}(t)\rho_k(x) \quad \text{in } L^2(0, T; H^2(\Omega))$$

and $\partial_t^\alpha(u^\varepsilon_N - a_N)$ in $L^2(0, T : L^2(\Omega))$, where $a_N = \sum_{k=1}^N c_k\rho_k$.

Using (4.21), multiplying the first equation in (4.20) by $d_t^\alpha p^\varepsilon_{N,\ell}$, summing over $\ell = 1, \ldots, N$, we obtain

$$(d_t^\alpha u^\varepsilon_N(\cdot, t), d_t^\alpha u^\varepsilon_N(\cdot, t))_{L^2(\Omega)} - \sum_{i,j=1}^n (\partial_i(a_{ij}(\cdot, t)\partial_j u^\varepsilon_N(\cdot, t)), d_t^\alpha u^\varepsilon_N(\cdot, t))_{L^2(\Omega)}$$

$$-\sum_{j=1}^n (b_j(\cdot, t)\partial_j u^\varepsilon_N(\cdot, t), d_t^\alpha u^\varepsilon_N(\cdot, t))_{L^2(\Omega)} - (c(\cdot, t)u^\varepsilon_N(\cdot, t), d_t^\alpha u^\varepsilon_N(\cdot, t))_{L^2(\Omega)}$$

$$= (F^\varepsilon_N(\cdot, t), d_t^\alpha u^\varepsilon_N(\cdot, t))_{L^2(\Omega)}. \tag{4.35}$$

The main part of the proof is the estimation of the second term on the left-hand side. By $u^\varepsilon_N(\cdot, t) \in H^1_0(\Omega)$, integration by parts yields

$$-\left(\sum_{i,j=1}^n \partial_i(a_{ij}(\cdot, t)\partial_j u^\varepsilon_N(\cdot, t)), d_t^\alpha u^\varepsilon_N(\cdot, t)\right)_{L^2(\Omega)}$$

$$= \left(\sum_{i,j=1}^n a_{ij}(\cdot, t)\partial_j u^\varepsilon_N(\cdot, t), d_t^\alpha \partial_i u^\varepsilon_N(\cdot, t)\right)_{L^2(\Omega)}.$$

Now we consider $\sum_{i,j=1}^n a_{ij}(\partial_j u^\varepsilon_N)\partial_i d_t^\alpha u^\varepsilon_N(\cdot, t)$. Since $a_{ij} = a_{ji}$, $1 \le i, j \le n$, we have

$$\sum_{i,j=1}^n a_{ij}(\partial_j u^\varepsilon_N)\partial_i d_t^\alpha u^\varepsilon_N(\cdot, t)$$

$$= \sum_{i>j} a_{ij}((\partial_j u^\varepsilon_N)\partial_i d_t^\alpha u^\varepsilon_N + (\partial_i u^\varepsilon_N)\partial_j d_t^\alpha u^\varepsilon_N) + \sum_{i-1}^n a_{ii}(\partial_i u^\varepsilon_N)(\partial_i d_t^\alpha u^\varepsilon_N) =: J_1 + J_2.$$

We set $g(t) = \frac{1}{\Gamma(1-\alpha)} t^{-\alpha}$. Here we write $u_N^\varepsilon(s) = u_N^\varepsilon(x, s)$, etc. Then

$$J_1 = \sum_{i>j} a_{ij}((\partial_j u_N^\varepsilon)\partial_i d_t^\alpha u_N^\varepsilon + (\partial_i u_N^\varepsilon)\partial_j d_t^\alpha u_N^\varepsilon)(x, t)$$

$$= \sum_{i>j} a_{ij}(x, t) \int_0^t g(t-s)((\partial_j u_N^\varepsilon)(t)\partial_i \partial_s u_N^\varepsilon(x, s) + (\partial_i u_N^\varepsilon)(t)\partial_j \partial_s u_N^\varepsilon(x, s))ds$$

$$= -\sum_{i>j} a_{ij}(x, t) \int_0^t g(t-s)((\partial_i \partial_s u_N^\varepsilon(s))(\partial_j u_N^\varepsilon(s) - \partial_j u_N^\varepsilon(t))$$

$$+ (\partial_j \partial_s u_N^\varepsilon(s))(\partial_i u_N^\varepsilon(s) - \partial_i u_N^\varepsilon(t))ds$$

$$+ \sum_{i>j} a_{ij}(x, t) \int_0^t g(t-s)((\partial_i \partial_s u_N^\varepsilon(s))\partial_j u_N^\varepsilon(s) + (\partial_j \partial_s u_N^\varepsilon(s))\partial_i u_N^\varepsilon(s))ds$$

$$=: J_{11} + J_{12}.$$

We have

$$J_{11} = -\sum_{i>j} a_{ij}(x, t) \int_0^t g(t-s)\{\partial_s(\partial_i u_N^\varepsilon(s) - \partial_i u_N^\varepsilon(t))(\partial_j u_N^\varepsilon(s) - \partial_j u_N^\varepsilon(t))$$

$$+ \partial_s(\partial_j u_N^\varepsilon(s) - \partial_j u_N^\varepsilon(t))(\partial_i u_N^\varepsilon(s) - \partial_i u_N^\varepsilon(t))\}ds$$

$$= -\sum_{i>j} a_{ij}(x, t) \int_0^t g(t-s)\partial_s((\partial_i u_N^\varepsilon(s) - \partial_i u_N^\varepsilon(t))(\partial_j u_N^\varepsilon(s) - \partial_j u_N^\varepsilon(t)))ds$$

$$= -\sum_{i>j} a_{ij}(x, t)[g(t-s)(\partial_i u_N^\varepsilon(s) - \partial_i u_N^\varepsilon(t))(\partial_j u_N^\varepsilon(s) - \partial_j u_N^\varepsilon(t))]_{s=0}^{s=t}$$

$$- \sum_{i>j} a_{ij}(x, t) \int_0^t \frac{dg}{d\xi}(t-s)(\partial_i u_N^\varepsilon(s) - \partial_i u_N^\varepsilon(t))(\partial_j u_N^\varepsilon(s) - \partial_j u_N^\varepsilon(t))ds$$

$$= \sum_{i>j} a_{ij}(x, t)g(t)(\partial_i u_N^\varepsilon(0) - \partial_i u_N^\varepsilon(t))(\partial_j u_N^\varepsilon(0) - \partial_j u_N^\varepsilon(t))$$

$$+ \int_0^t -\frac{dg}{d\xi}(t-s) \sum_{i>j} a_{ij}(x, t)(\partial_i u_N^\varepsilon(s) - \partial_i u_N^\varepsilon(t))(\partial_j u_N^\varepsilon(s) - \partial_j u_N^\varepsilon(t))ds.$$

Here we used

$$\lim_{s \uparrow t} g(t-s)(\partial_i u_N^\varepsilon(s) - \partial_i u_N^\varepsilon(t))(\partial_j u_N^\varepsilon(s) - \partial_j u_N^\varepsilon(t)) = 0,$$

which can be justified by $\partial_i u_N^\varepsilon(x, \cdot)$, $\partial_j u_N^\varepsilon(x, \cdot) \in W_\alpha(0, T)$ following from Theorem 3.6 (v).

On the other hand,

$$J_{12} = \sum_{i>j} a_{ij}(x,t) \int_0^t g(t-s) \partial_s (\partial_i u_N^\varepsilon(x,s) \partial_j u_N^\varepsilon(x,s)) ds$$

$$= \sum_{i>j} a_{ij}(x,t) d_t^\alpha (\partial_i u_N^\varepsilon \partial_j u_N^\varepsilon)(x,t).$$

Consequently,

$$J_1 = \frac{1}{2} \sum_{i \neq j} a_{ij} g(t)(\partial_i u_N^\varepsilon(0) - \partial_i u_N^\varepsilon(t))(\partial_j u_N^\varepsilon(0) - \partial_j u_N^\varepsilon(t))$$

$$+ \frac{1}{2} \int_0^t -\frac{dg}{d\xi}(t-s) \sum_{i \neq j} a_{ij}(x,t)(\partial_i u_N^\varepsilon(s) - \partial_i u_N^\varepsilon(t))(\partial_j u_N^\varepsilon(s) - \partial_j u_N^\varepsilon(t)) ds$$

$$+ \frac{1}{2} \sum_{i \neq j} a_{ij} d_t^\alpha (\partial_i u_N^\varepsilon \partial_j u_N^\varepsilon)(x,t).$$

Similarly, we can calculate

$$J_2 = \sum_i a_{ii} \int_0^t g(t-s) \partial_i \partial_s u_N^\varepsilon(s) \partial_i u_N^\varepsilon(t) ds$$

$$= -\sum_i a_{ii} \int_0^t g(t-s) \partial_i \partial_s u_N^\varepsilon(s)(\partial_i u_N^\varepsilon(s) - \partial_i u_N^\varepsilon(t)) ds$$

$$+ \sum_i a_{ii} \int_0^t g(t-s) \partial_i \partial_s u_N^\varepsilon(s) \partial_i u_N^\varepsilon(s) ds$$

$$= -\frac{1}{2} \sum_i a_{ii} \int_0^t g(t-s) \partial_s ((\partial_i u_N^\varepsilon(s) - \partial_i u_N^\varepsilon(t))(\partial_i u_N^\varepsilon(s) - \partial_i u_N^\varepsilon(t))) ds$$

$$+ \frac{1}{2} \sum_i a_{ii} \int_0^t g(t-s) \partial_s (|\partial_i u_N^\varepsilon(s)|^2) ds$$

$$= \frac{1}{2} \sum_i a_{ii} g(t)(\partial_i u_N^\varepsilon(t) - \partial_i u_N^\varepsilon(0))^2 + \frac{1}{2} \int_0^t -\frac{dg}{d\xi}(t-s) \sum_i a_{ii}(\partial_i u_N^\varepsilon(s) - \partial_i u_N^\varepsilon(t))^2 ds$$

$$+ \frac{1}{2} \sum_i a_{ii} d_t^\alpha (|\partial_i u_N^\varepsilon(t)|^2).$$

Thus

$$\sum_{i,j=1}^{n} a_{ij}\partial_j u_N^\varepsilon(x,t)\partial_i d_t^\alpha u_N^\varepsilon(x,t)$$

$$=\frac{1}{2}\sum_{i,j=1}^{n} g(t)a_{ij}(x,t)(\partial_i u_N^\varepsilon(0)-\partial_i u_N^\varepsilon(t))(\partial_j u_N^\varepsilon(0)-\partial_j u_N^\varepsilon(t))$$

$$+\frac{1}{2}\int_0^t -\frac{dg}{d\xi}(t-s)\sum_{i,j=1}^{n} a_{ij}(x,t)(\partial_i u_N^\varepsilon(s)-\partial_i u_N^\varepsilon(t))(\partial_j u_N^\varepsilon(s)-\partial_j u_N^\varepsilon(t))ds$$

$$+\frac{1}{2}\sum_{i,j=1}^{n} a_{ij}d_t^\alpha((\partial_i u_N^\varepsilon)\partial_j u_N^\varepsilon)(x,t).$$

By $g(t)>0$, $-\frac{dg}{d\xi}(t-s)>0$ for $0<s<t<T$ and (4.2), we obtain

$$2\sum_{i,j=1}^{n} a_{ij}(x,t)\partial_j u_N^\varepsilon(x,t)\partial_i d_t^\alpha u_N^\varepsilon(x,t) \geq \sum_{i,j=1}^{n} a_{ij}(x,t)d_t^\alpha((\partial_i u_N^\varepsilon)\partial_j u_N^\varepsilon)(x,t)$$

$$=\int_0^t g(t-s)\sum_{i,j=1}^{n} \partial_s(a_{ij}(x,s)\partial_i u_N^\varepsilon(s)\partial_j u_N^\varepsilon(s))ds$$

$$-\int_0^t g(t-s)\sum_{i,j=1}^{n} (\partial_s a_{ij}(x,s))\partial_i u_N^\varepsilon(s)\partial_j u_N^\varepsilon(s)ds$$

$$+\int_0^t g(t-s)\sum_{i,j=1}^{n} (a_{ij}(x,t)-a_{ij}(x,s))\partial_s((\partial_i u_N^\varepsilon(s))\partial_j u_N^\varepsilon(s))ds =: S_1 + S_2 + S_3.$$

$$(4.36)$$

Here we have

$$S_1 = \int_0^t g(t-s)\sum_{i,j=1}^{n} \partial_s(a_{ij}(x,s)\partial_i u_N^\varepsilon(s)\partial_j u_N^\varepsilon(s))ds \qquad (4.37)$$

$$=\int_0^t g(s)\partial_t\left(\sum_{i,j=1}^{n} a_{ij}(x,t-s)\partial_i u_N^\varepsilon(t-s)\partial_j u_N^\varepsilon(t-s)\right)ds$$

$$=\partial_t\int_0^t g(s)\sum_{i,j=1}^{n} a_{ij}(x,t-s)\partial_i u_N^\varepsilon(t-s)\partial_j u_N^\varepsilon(t-s)ds$$

$$-g(t)\sum_{i,j=1}^{n} a_{ij}(x,0)\partial_i u_N^\varepsilon(x,0)\partial_j u_N^\varepsilon(x,0).$$

Furthermore

$$S_2 = -\int_0^t g(t-s) \sum_{i,j=1}^n (\partial_s a_{ij}(x,s)) \partial_i u_N^\varepsilon(s) \partial_j u_N^\varepsilon(s) ds$$

and

$$S_3 = \int_0^t \sum_{i,j=1}^n r_{ij}(x,t,s) \partial_s (\partial_i u_N^\varepsilon(s) \partial_j u_N^\varepsilon(s)) ds,$$

where

$$r_{ij}(x,t,s) = \frac{(t-s)^{-\alpha}}{\Gamma(1-\alpha)} (a_{ij}(x,t) - a_{ij}(x,s)), \quad 1 \le i,j \le n.$$

We directly obtain

$$|S_2| \le C \int_0^t (t-s)^{-\alpha} |\nabla u_N^\varepsilon(\cdot,s)|^2 ds. \tag{4.38}$$

To estimate $|S_3|$, we integrate by parts and apply

$$|\partial_s r_{ij}(x,t,s)| = \left| \frac{\alpha(t-s)^{-\alpha-1}}{\Gamma(1-\alpha)} (a_{ij}(x,t) - a_{ij}(x,s)) - \frac{(t-s)^{-\alpha}}{\Gamma(1-\alpha)} \partial_s a_{ij}(x,s) \right|$$

$$\le C|t-s|^{-\alpha}$$

by $a_{ij} \in C^1([0,T]; C(\overline{\Omega}))$, so that

$$|S_3| = \left| \left[\sum_{i,j=1}^n r_{ij}(x,t,s) \partial_i u_N^\varepsilon(s) \partial_j u_N^\varepsilon(s) \right]_{s=0}^{s=t} \right.$$

$$\left. - \int_0^t \sum_{i,j=1}^n (\partial_s r_{ij})(x,t,s) \partial_i u_N^\varepsilon(s) \partial_j u_N^\varepsilon(s) ds \right|$$

$$\le C|\nabla u_N^\varepsilon(x,0)|^2 + C \int_0^t (t-s)^{-\alpha} |\nabla u_N^\varepsilon(x,s)|^2 ds.$$

Therefore (4.36)–(4.38) imply

$$2 \sum_{i,j=1}^{n} a_{ij}(x,t)\partial_j u_N^\varepsilon(x,t)\partial_i d_t^\alpha u_N^\varepsilon(x,t)$$

$$\geq \partial_t \int_0^t g(s) \sum_{i,j=1}^{n} a_{ij}(x,t-s)\partial_i u_N^\varepsilon(x,t-s)\partial_j u_N^\varepsilon(x,t-s)ds$$

$$- g(t) \sum_{i,j=1}^{n} a_{ij}(x,0)\partial_i u_N^\varepsilon(x,0)\partial_j u_N^\varepsilon(x,0)$$

$$- C|\nabla u_N^\varepsilon(x,0)|^2 - C\int_0^t (t-s)^{-\alpha}|\nabla u_N^\varepsilon(x,s)|^2 ds, \quad x \in \Omega, 0 < t < T.$$

$$(4.39)$$

Taking $\int_0^T \int_\Omega \cdots dx dt$ and applying the Young inequality (Lemma A.1 in the Appendix) to the fourth term on the right-hand side of (4.39), we have

$$2 \int_0^T \sum_{i,j=1}^{n} (a_{ij}(\cdot,t)\partial_j u_N^\varepsilon(\cdot,t), d_t^\alpha \partial_i u_N^\varepsilon(\cdot,t))_{L^2(\Omega)}dt$$

$$\geq \int_0^T g(s) \sum_{i,j=1}^{n} (a_{ij}(T-s)\partial_i u_N^\varepsilon(T-s), \partial_j u_N^\varepsilon(T-s))_{L^2(\Omega)}ds$$

$$- C\int_0^T g(t)dt\|\nabla a\|_{L^2(\Omega)}^2 - C(\|\nabla a\|_{L^2(\Omega)} + \|\nabla u_N^\varepsilon\|_{L^2(0,T;L^2(\Omega))}^2).$$

Since

$$\int_0^T g(s) \sum_{i,j=1}^{n} (a_{ij}(\cdot,T-s)\partial_i u_N^\varepsilon(\cdot,T-s), \partial_j u_N^\varepsilon(\cdot,T-s))_{L^2(\Omega)}ds$$

$$= \int_0^T g(T-s) \sum_{i,j=1}^{n} (a_{ij}(\cdot,s)\partial_i u_N^\varepsilon(\cdot,s), \partial_j u_N^\varepsilon(\cdot,s))_{L^2(\Omega)}ds$$

$$\geq CT^{-\alpha}\|\nabla u_N^\varepsilon\|_{L^2(0,T;L^2(\Omega))}^2$$

by (4.2), we obtain

$$\int_0^T [\text{the second term on the left-hand side of (4.35)}]dt \qquad (4.40)$$

$$= \int_0^T \left(\sum_{i,j=1}^n a_{ij}(\cdot, t) \partial_j u_N^\varepsilon(\cdot, t), \, d_t^\alpha \partial_i u_N^\varepsilon(\cdot, t) \right)_{L^2(\Omega)} dt$$

$$\geq - C(\|\nabla u_N^\varepsilon\|_{L^2(0,T;L^2(\Omega))}^2 + \|\nabla a\|_{L^2(\Omega)}^2).$$

By the Poincaré inequality and $u_N^\varepsilon(\cdot, t) \in H_0^1(\Omega)$, for small $\delta > 0$, we can estimate

$$\int_0^T |[\text{the third and the fourth terms on the left-hand side of (4.35)}]dt$$

$$\leq \int_0^T (C_\delta \|\nabla u_N^\varepsilon(\cdot, t)\|_{L^2(\Omega)}^2 + \delta \|d_t^\alpha u_N^\varepsilon(\cdot, t)\|_{L^2(\Omega)}^2)dt \qquad (4.41)$$

and

$$\left| \int_0^T (F_N^\varepsilon(\cdot, t), d_t^\alpha u_N^\varepsilon(\cdot, t))_{L^2(\Omega)}dt \right|$$

$$\leq C_\delta \|F_N^\varepsilon\|_{L^2(0,T;L^2(\Omega))}^2 + C\delta \|d_t^\alpha u_N^\varepsilon(\cdot, t)\|_{L^2(0,T;L^2(\Omega))}^2. \qquad (4.42)$$

Hence, applying (4.40)–(4.42) in (4.35), we derive

$$\|d_t^\alpha u_N^\varepsilon\|_{L^2(0,T;L^2(\Omega))}^2 \leq C\|\nabla a\|_{L^2(\Omega)}^2 + C_\delta \|\nabla u_N^\varepsilon\|_{L^2(0,T;L^2(\Omega))}^2$$

$$+ \delta \|d_t^\alpha u_N^\varepsilon\|_{L^2(0,T;L^2(\Omega))}^2 + C_\delta \|F_N^\varepsilon\|_{L^2(0,T;L^2(\Omega))}^2.$$

Fixing $\delta > 0$ sufficiently small and absorbing the third term on the right-hand side into the left-hand side and applying (4.30), which was verified in the proof of Theorem 4.1, we obtain

$$\|\partial_t^\alpha (u_N^\varepsilon - a)\|_{L^2(0,T;L^2(\Omega))}^2 = \|d_t^\alpha u_N^\varepsilon\|_{L^2(0,T;L^2(\Omega))}^2$$

$$\leq C(\|\nabla a_N\|_{L^2(\Omega)}^2 + \|F_N^\varepsilon\|_{L^2(0,T;L^2(\Omega))}^2), \qquad (4.43)$$

where the constant $C > 0$ is independent of $N \in \mathbb{N}$ and $\varepsilon > 0$.

Based on the uniform bound (4.43) in N and ε, we can argue similarly to (4.31)–(4.33). We omit the details, so that we can complete the proof of Theorem 4.2.

We note that the uniqueness of solutions is proved for less regular class, and the uniqueness holds also within the stronger regularity of solutions in Theorem 4.2.

4.5 Proofs of Propositions 4.1 and 4.2

4.5.1 Proof of Proposition 4.1

As is already proved, $u \in H_\alpha(0, T; H^{-1}(\Omega)) \cap L^2(0, T; H_0^1(\Omega))$ satisfies

$$\begin{cases} \partial_t^\alpha (u - a) + A(t)u = F & \text{in } H^{-1}(\Omega), 0 < t < T, \\ u(\cdot, t) \in H_0^1(\Omega), \quad u - a \in H_\alpha(0, T; H^{-1}(\Omega)). \end{cases}$$

Then we have

$$_{H^{-1}(\Omega)}\langle \partial_s^\alpha (u - a)(\cdot, s), u - a \rangle_{H_0^1(\Omega)} + {}_{H^{-1}(\Omega)}\langle A(s)(u - a)(\cdot, s), u - a \rangle_{H_0^1(\Omega)}$$

$$+ {}_{H^{-1}(\Omega)}\langle A(s)a, u - a \rangle_{H_0^1(\Omega)} = {}_{H^{-1}(\Omega)}\langle F(\cdot, s), (u - a)(\cdot, s) \rangle_{H_0^1(\Omega)}$$

for $0 < s < T$. Here we note that for any $\varepsilon > 0$ there exists a constant $C_\varepsilon > 0$ such that

$$\begin{cases} |{}_{H^{-1}(\Omega)}\langle A(s)a, u(\cdot, s) - a \rangle_{H_0^1(\Omega)}| \le C\|a\|_{H_0^1(\Omega)} \|(u - a)(\cdot, s)\|_{H_0^1(\Omega)} \\ \le \varepsilon \|(u - a)(\cdot, s)\|_{H_0^1(\Omega)}^2 + C_\varepsilon \|a\|_{H_0^1(\Omega)}^2, \\ \left| \sum_{j=1}^n {}_{H^{-1}(\Omega)}\langle b_j(\cdot, s)\partial_j(u(\cdot, s) - a), u(\cdot, s) - a \rangle_{H_0^1(\Omega)} \right| \\ + |{}_{H^{-1}(\Omega)}\langle c(\cdot, s)(u(\cdot, s) - a), u(\cdot, s) - a \rangle_{H_0^1(\Omega)}| \\ \le C\|u(\cdot, s) - a\|_{L^2(\Omega)} \|u(\cdot, s) - a\|_{H_0^1(\Omega)} \\ \le \varepsilon \|u(\cdot, s) - a\|_{H_0^1(\Omega)}^2 + C_\varepsilon \|u(\cdot, s)0 - a\|_{L^2(\Omega)}^2. \end{cases}$$

$$(4.44)$$

Here we used

$$\|b_j(\cdot, s)\partial_j(u(\cdot, s) - a)\|_{H^{-1}(\Omega)} = \sup_{\|\psi\|_{H_0^1(\Omega)} = 1} |{}_{H^{-1}(\Omega)}\langle b_j(\cdot, s)\partial_j(u(\cdot, s) - a), \psi \rangle_{H_0^1(\Omega)}|$$

$$= \sup_{\|\psi\|_{H_0^1(\Omega)} = 1} |(u(\cdot, s) - a, \partial_j(b_j\psi))_{L^2(\Omega)}|$$

$$\le \sup_{\|\psi\|_{H_0^1(\Omega)} = 1} \|u(\cdot, s) - a\|_{L^2(\Omega)} \|\partial_j(b_j\psi)\|_{L^2(\Omega)} \le C\|u(\cdot, s) - a\|_{L^2(\Omega)}.$$

Multiplying $\frac{(t-s)^{\alpha-1}}{\Gamma(\alpha)}$ and integrating in s over $(0, t)$ and applying (4.2) and Theorem 3.4 (ii), we obtain

$$\|u(\cdot, t) - a\|_{L^2(\Omega)}^2 + \int_0^t (t-s)^{\alpha-1} \|\nabla(u(\cdot, s) - a)\|_{L^2(\Omega)}^2 ds$$

$$\leq C \int_0^t (t-s)^{\alpha-1} |_{H^{-1}(\Omega)} \langle A(s)a, u(\cdot, s) - a \rangle_{H_0^1(\Omega)}| ds$$

$$+ \left| \int_0^t (t-s)^{\alpha-1} \sum_{j=1}^n {}_{H^{-1}(\Omega)} \langle b_j(\cdot, s) \partial_j (u(\cdot, s) - a), u(\cdot, s) - a \rangle_{H_0^1(\Omega)} ds \right|$$

$$+ \left| \int_0^t {}_{H^{-1}(\Omega)} \langle c(\cdot, s)(u(\cdot, s) - a), u(\cdot, s) - a \rangle_{H_0^1(\Omega)} (t-s)^{\alpha-1} ds \right|$$

$$+ C \int_0^t (t-s)^{\alpha-1} \|F(\cdot, s)\|_{H^{-1}(\Omega)} \|u(\cdot, s) - a\|_{H_0^1(\Omega)} ds.$$

Applying (4.44) and

$$\|F(\cdot, s)\|_{H^{-1}(\Omega)} \|u(\cdot, s) - a\|_{H_0^1(\Omega)} \leq \varepsilon \|u(\cdot, s) - a\|_{H_0^1(\Omega)}^2 + C_\varepsilon \|F(\cdot, s)\|_{H^{-1}(\Omega)}^2,$$

we obtain

$$\|u(\cdot, t) - a\|_{L^2(\Omega)}^2 + \int_0^t (t-s)^{\alpha-1} \|u(\cdot, s) - a\|_{H_0^1(\Omega)}^2 ds$$

$$\leq \varepsilon \int_0^t (t-s)^{\alpha-1} \|u(\cdot, s) - a\|_{H_0^1(\Omega)}^2 ds + C_\varepsilon \int_0^t (t-s)^{\alpha-1} \|u(\cdot, s) - a\|_{L^2(\Omega)}^2 ds$$

$$+ C_\varepsilon \int_0^t (t-s)^{\alpha-1} \|a\|_{H_0^1(\Omega)}^2 ds + C_\varepsilon \int_0^t (t-s)^{\alpha-1} \|F(\cdot, s)\|_{H^{-1}(\Omega)}^2 ds.$$

Choosing $\varepsilon > 0$ sufficiently small, we absorb the first term on the right-hand side into the left-hand side, so that

$$\|u(\cdot, t) - a\|_{L^2(\Omega)}^2$$

$$\leq \|u(\cdot, t) - a\|_{L^2(\Omega)}^2 + C \int_0^t (t-s)^{\alpha-1} \|u(\cdot, s) - a\|_{H_0^1(\Omega)}^2 ds$$

$$\leq C \int_0^t (t-s)^{\alpha-1} \|u(\cdot, s) - a\|_{L^2(\Omega)}^2 ds$$

$$+ Ct^\alpha \|a\|_{H_0^1(\Omega)}^2 + C \int_0^t (t-s)^{\alpha-1} \|F(\cdot, s)\|_{H^{-1}(\Omega)}^2 ds.$$

Since $F \in L^p(0, T; H^{-1}(\Omega))$ with $p > \frac{2}{\alpha}$, we can choose $q > 1$ satisfying $\frac{1}{2} + \frac{1}{q} = 1$ and $q(\alpha - 1) > -1$. Indeed, $q = \frac{p}{p-2}$ and $p > \frac{2}{\alpha}$ imply $q(\alpha - 1) = \frac{p}{p-2}(\alpha - 1) > -1$. Therefore the Hölder inequality yields

$$\int_0^t (t-s)^{\alpha-1} \|F(\cdot, s)\|_{H^{-1}(\Omega)}^2 ds \leq \left(\int_0^t (t-s)^{q\alpha-q} ds \right)^{\frac{1}{q}} \left(\int_0^t (\|F(\cdot, s)\|_{H^{-1}(\Omega)}^2)^{\frac{p}{2}} ds \right)^{\frac{2}{p}}$$

$$\leq C t^{\frac{\theta}{q}} \|F\|_{L^p(0,t;H^{-1}(\Omega))}^2,$$

where $\theta = q(\alpha - 1) + 1 > 0$. Thus

$$\|u(\cdot, t) - a\|_{L^2(\Omega)}^2 \leq C(t^\alpha \|a\|_{H_0^1(\Omega)}^2 + t^{\frac{\theta}{q}} \|F\|_{L^p(0,T;H^{-1}(\Omega))}^2)$$

$$+ C \int_0^t (t-s)^{\alpha-1} \|u(\cdot, s) - a\|_{L^2(\Omega)}^2 ds, \quad 0 < t < T.$$

Lemma A.2 in the Appendix implies

$$\|u(\cdot, t) - a\|_{L^2(\Omega)}^2 \leq C((t^\alpha + t^{2\alpha}) \|a\|_{H_0^1(\Omega)}^2 + (t^{\frac{\theta}{q}} + t^{\alpha + \frac{\theta}{q}}) \|F\|_{L^p(0,T;H^{-1}(\Omega))}^2),$$

which completes the proof of Proposition 4.1.

4.5.2 Proof of Proposition 4.2

The proof is based on an estimate similar to Proposition 4.1.

By (4.3), we have

$$\partial_s^\alpha (u - a)(\cdot, s) + A(s)u(x, s) = 0 \quad \text{in } L^2(\Omega), 0 < s < t.$$

Multiplying both sides by $(u - a)(x, s)(t - s)^{\alpha-1}$ and integrating over $\Omega \times (0, t)$, we obtain

$$\int_0^t \left(\int_\Omega (\partial_s^\alpha (u-a)(x,s))(u-a)(x,s)dx \right) (t-s)^{\alpha-1} ds$$

$$- \int_0^t \left(\int_\Omega \sum_{i,j=1}^n \partial_i (a_{ij}(x,s)\partial_j u(x,s))u(x,s)dx \right) (t-s)^{\alpha-1} ds$$

$$+ \int_0^t \left(\int_\Omega \sum_{i,j=1}^n \partial_i (a_{ij}(x,s)\partial_j u(x,s))a(x)dx \right) (t-s)^{\alpha-1} ds$$

$$-\int_0^t \left(\int_\Omega \sum_{j=1}^n b_j(x,s) \partial_j u(x,s)(u(x,s) - a(x)) dx \right) (t-s)^{\alpha-1} ds$$

$$-\int_0^t \left(\int_\Omega c(x,s) u(x,s)(u(x,s) - a(x)) dx \right) (t-s)^{\alpha-1} ds = 0.$$

Since $(u-a)(x,\cdot) \in H_\alpha(0,T)$ for $x \in \Omega$ and $u(\cdot,s) \in H_0^1(\Omega)$ for $0 < s < t$, we integrate by parts and use Theorem 3.4 (ii) where we note

$$_{H^{-1}(\Omega)} \langle \partial_s^\alpha(u(\cdot,s)-a), u(\cdot,s)-a \rangle_{H_0^1(\Omega)} = \int_\Omega (\partial_s^\alpha(u(x,s)-a(x))(u(x,s)-a(x)) dx,$$

by $\partial_s^\alpha(u-a) \in L^2(0,T; L^2(\Omega))$, so that

$$C_0 \|u(\cdot,t) - a\|_{L^2(\Omega)}^2 + \int_0^t \left(\int_\Omega \sum_{i,j=1}^n a_{ij}(x,s) \partial_j u(x,s) \partial_i u(x,s) dx \right) (t-s)^{\alpha-1} ds$$

$$-\int_0^t \left(\int_\Omega \sum_{i,j=1}^n a_{ij}(x,s) \partial_j u(x,s) \partial_i a(x) dx \right) (t-s)^{\alpha-1} ds$$

$$-\int_0^t \left(\int_\Omega \sum_{j=1}^n b_j(x,s) \partial_j u(x,s) u(x,s) dx \right) (t-s)^{\alpha-1} ds$$

$$+\int_0^t \left(\int_\Omega \sum_{j=1}^n b_j(x,s) \partial_j u(x,s) a(x) dx \right) (t-s)^{\alpha-1} ds$$

$$-\int_0^t \left(\int_\Omega c(x,s) u(x,s)^2 dx \right) (t-s)^{\alpha-1} ds$$

$$+\int_0^t \left(\int_\Omega c(x,s) u(x,s) a(x) dx \right) (t-s)^{\alpha-1} ds = 0.$$

By (4.2) and using $|u||a| \leq \frac{1}{2}(|u|^2 + |a|^2)$, we obtain

$$C_0 \|u(\cdot,t)\|_{L^2(\Omega)}^2 - C_1 \|a\|_{L^2(\Omega)}^2 + C_2 \int_0^t \left(\int_\Omega |\nabla u(x,t)|^2 (t-s)^{\alpha-1} dx \right) ds$$

$$-\int_0^t \int_\Omega C_3 \varepsilon |\nabla u(x,s)|(t-s)^{\frac{\alpha-1}{2}} \left(\frac{1}{\varepsilon} |\nabla a(x)|(t-s)^{\frac{\alpha-1}{2}} \right) dx ds$$

$$-\int_0^t \int_\Omega C_3 \varepsilon |\nabla u(x,s)|(t-s)^{\frac{\alpha-1}{2}} \left(\frac{1}{\varepsilon} |u(x,s)|(t-s)^{\frac{\alpha-1}{2}} \right) dx ds$$

$$-\int_0^t \int_\Omega C_3 \varepsilon |\nabla u(x,s)| (t-s)^{\frac{\alpha-1}{2}} \left(\frac{1}{\varepsilon} |a(x)| (t-s)^{\frac{\alpha-1}{2}} \right) dxds$$

$$-C_3 \int_0^t \int_\Omega |u(x,s)|^2 (t-s)^{\alpha-1} dxds - C_3 \int_0^t \int_\Omega |a(x)|^2 (t-s)^{\alpha-1} dxds \leq 0.$$

Hence, after applying the Cauchy–Schwarz inequality, choosing $\varepsilon > 0$ sufficiently small and absorbing $\int_0^t \int_\Omega C_3^2 \varepsilon^2 |\nabla u(x,s)|^2 (t-s)^{\alpha-1} dxds$ into the term $C_2 \int_0^t \int_\Omega |\nabla u(x,s)|^2 (t-s)^{\alpha-1} dxds$, we obtain

$$\|u(\cdot,t)\|^2_{L^2(\Omega)} + C_4 \int_0^t \int_\Omega |\nabla u(x,t)|^2 (t-s)^{\alpha-1} dxds$$

$$\leq C \int_0^t \left(\int_\Omega (|\nabla a(x)|^2 + |a(x)|^2) dx \right) (t-s)^{\alpha-1} dxds$$

$$+ C \int_0^t \int_\Omega |u(x,s)|^2 (t-s)^{\alpha-1} dxds, \quad t > 0.$$

Therefore

$$\|u(\cdot,t)\|^2_{L^2(\Omega)} + \int_0^t \|u(\cdot,s)\|^2_{H_0^1(\Omega)} (t-s)^{\alpha-1} ds$$

$$\leq Ct^\alpha \|a\|^2_{H_0^1(\Omega)} + C \int_0^t \|u(\cdot,s)\|^2_{L^2(\Omega)} (t-s)^{\alpha-1} ds, \quad t > 0,$$

that is,

$$\|u(\cdot,t)\|^2_{L^2(\Omega)} \leq Ct^\alpha \|a\|^2_{H_0^1(\Omega)} + C \int_0^t \|u(\cdot,s)\|^2_{L^2(\Omega)} (t-s)^{\alpha-1} ds, \quad t > 0.$$

Consequently, noting that the constant $C > 0$ can be chosen uniformly in all $t > 0$ and applying Lemma A.2 in the Appendix, we can take a constant $C_0 > 0$ such that

$$\|u(\cdot,t)\|^2_{L^2(\Omega)} \leq C_0 t^\alpha \|a\|^2_{H_0^1(\Omega)} + C_0 e^{C_0 t} \left(\int_0^t (t-s)^{\alpha-1} s^\alpha ds \right) \|a\|^2_{H_0^1(\Omega)}$$

$$\leq \left(C_0 t^\alpha + C_0 e^{C_0 t} \frac{\Gamma(\alpha)\Gamma(\alpha+1)}{\Gamma(2\alpha+1)} t^{2\alpha} \right) \|a\|^2_{H_0^1(\Omega)}$$

for all $t \geq 0$. Thus the proof of Proposition 4.2 is complete.

4.6 Case Where the Coefficients of $A(t)$ Are Independent of Time

In the case where the coefficients of $A(t)$ are independent of t, we can represent the solution to the initial boundary value problem by the Mittag-Lefller functions (e.g., Sakamoto and Yamamoto [26]). In this section, we show that such a represented solution coincides with the solution established by Theorem 4.1.

Let

$$-Av(x) = \sum_{i,j=1}^{n} \partial_i(a_{ij}(x)\partial_j v(x)) + c(x)v(x), \quad x \in \Omega,$$

where $a_{ij} = a_{ji} \in C^1(\overline{\Omega})$, $1 \le i, j \le n$, $c \in L^\infty(\Omega)$, $c \le 0$ in Ω, and (4.2) is assumed to hold. We consider the operator A with the domain $\mathcal{D}(A) = H^2(\Omega) \cap H_0^1(\Omega)$. Then it is known that there exist eigenvalues of A which can be numbered with their multiplicities

$$0 < \lambda_1 \le \lambda_2 \le \cdots \longrightarrow \infty.$$

Henceforth if there is no fear of confusion, then we write the scalar product (f, g) for $f, g \in L^2(\Omega)$, in place of $(f, g)_{L^2(\Omega)}$.

Let φ_k, $k \in \mathbb{N}$ be an eigenfunction of A for λ_n such that

$$(\varphi_k, \varphi_\ell) := \int_\Omega \varphi_k(x)\varphi_\ell(x)dx = \begin{cases} 1, \ k = \ell, \\ 0, \ k \ne \ell. \end{cases}$$

Moreover, it is known that φ_k, $k \in \mathbb{N}$, is an orthonormal basis in $L^2(\Omega)$.

We can define a fractional power A^γ for $\gamma \in \mathbb{R}$ (e.g., Pazy [23]), and for $a \in \mathcal{D}(A^\gamma)$, we have

$$A^\gamma a = \sum_{k=1}^{\infty} \lambda_k^\gamma (a, \varphi_k)\varphi_k$$

where the series is convergent in $L^2(\Omega)$, and

$$\|A^\gamma u\|_{L^2(\Omega)} = \left(\sum_{k=1}^{\infty} \lambda_k^{2\gamma} (a, \varphi_k)^2 \right)^{\frac{1}{2}}.$$

For $a \in L^2(\Omega)$ and $F \in L^2(0, T; H^{-1}(\Omega))$, we consider

$$v(x, t) = \sum_{k=1}^{\infty} (a, \varphi_k) E_{\alpha,1}(-\lambda_k t^\alpha) \varphi_k(x)$$

$$+ \sum_{k=1}^{\infty} \left(\int_0^t {}_{H^{-1}(\Omega)} \langle F(\cdot, s), \varphi_k \rangle_{H_0^1(\Omega)} (t-s)^{\alpha-1} E_{\alpha,\alpha}(-\lambda_k(t-s)^\alpha) ds \right) \varphi_k(x)$$

$$=: v_1(x, t) + v_2(x, t). \tag{4.45}$$

We will verify:

Theorem 4.3 *Let $a \in L^2(\Omega)$ and $F \in L^2(0, T; H^{-1}(\Omega))$. Then v given by (4.45) satisfies $v - a \in H_\alpha(0, T; H^{-1}(\Omega))$, $v \in L^2(0, T; H_0^1(\Omega))$ and (4.3)–(4.5).*

The theorem means that (4.45) coincides with the solution by Theorem 4.1. We can similarly prove that v defined by (4.45) is the same as u in Theorem 4.2 if $a \in H_0^1(\Omega)$ and $F \in L^2(0, T; L^2(\Omega))$, but we omit the proof.

Proof It is sufficient to prove that $v - a \in H_\alpha(0, T; H^{-1}(\Omega))$ and v satisfies (4.3).

First Step First we note that $\mathcal{D}(A^{-\frac{1}{2}}) = H^{-1}(\Omega) = (H_0^1(\Omega))'$ and there exist constants $C_1, C_2 > 0$ such that

$$C_1 \|w\|_{H^{-1}(\Omega)} \le \|A^{-\frac{1}{2}} w\|_{L^2(\Omega)} \le C_2 \|w\|_{H^{-1}(\Omega)}$$

for $w \in H^{-1}(\Omega)$, and

$$\|A^{-\frac{1}{2}} w\|^2 = \sum_{k=1}^{\infty} \lambda_k^{-1} (w, \varphi_k)^2.$$

Since $a = \sum_{k=1}^{\infty} (a, \varphi_k) \varphi_k$ in $L^2(\Omega)$, we have

$$v_1(x, t) - a(x) = \sum_{k=1}^{\infty} (a, \varphi_k)(E_{\alpha,1}(-\lambda_k t^\alpha) - 1) \varphi_k(x). \tag{4.46}$$

Second Step Theorem 2.4 implies

$$\|(v_1 - a)(x, \cdot)\|_{H_\alpha(0,T)} \sim \|\partial_t^\alpha (v_1 - a)(x, \cdot)\|_{L^2(0,T)}$$

for almost all $x \in \Omega$. Moreover, since $E_{\alpha,1}(-\lambda_k t^\alpha) - 1 \in H_\alpha(0, T) \cap {}_0W^{1,1}(0, T)$ by Proposition 2.2, we see that

$$d_t^\alpha(E_{\alpha,1}(-\lambda_k t^\alpha) - 1) = \partial_t^\alpha(E_{\alpha,1}(-\lambda_k t^\alpha) - 1),$$

so that

$$\partial_t^\alpha (v_1 - a)(x,t) = \sum_{k=1}^\infty (a, \varphi_k) \partial_t^\alpha (E_{\alpha,1}(-\lambda_k t^\alpha) - 1) \varphi_k(x)$$

$$= - \sum_{k=1}^\infty \lambda_k (a, \varphi_k) E_{\alpha,1}(-\lambda_k t^\alpha) \varphi_k(x). \qquad (4.47)$$

Therefore Lemma 2.5 (i) implies

$$\|\partial_t^\alpha (v_1 - a)(\cdot, t)\|_{H^{-1}(\Omega)}^2 = \sum_{k=1}^\infty (a, \varphi_k)^2 |E_{\alpha,1}(-\lambda_k t^\alpha)|^2 \lambda_k$$

$$\leq C t^{-\alpha} \sum_{k=1}^\infty (a, \varphi_k)^2 \left(\frac{\lambda_k^{\frac{1}{2}} t^{\frac{\alpha}{2}}}{1 + \lambda_k t^\alpha} \right)^2 \leq C t^{-\alpha} \|a\|_{L^2(\Omega)}^2,$$

that is,

$$\|\partial_t^\alpha (v_1 - a)\|_{L^2(0,T;H^{-1}(\Omega))}^2 \leq C \|a\|_{L^2(\Omega)}^2. \qquad (4.48)$$

Hence $v_1 - a \in H_\alpha(0, T; H^{-1}(\Omega))$.

Third Step   First we note an important property of $E_{\alpha,1}$, which is a consequence of Lemma 2.5 (ii) and what is called the complete monotonicity:

$$\frac{d}{d\eta} E_{\alpha,1}(-\eta^\alpha) \leq 0, \quad E_{\alpha,\alpha}(-\eta^\alpha) \geq 0, \quad \eta \geq 0, \quad 0 < \alpha < 1 \qquad (4.49)$$

(e.g., Gorenflo et al. [10], Miller and Samko [22]). Therefore we obtain

$$\int_0^\eta |s^{\alpha-1} E_{\alpha,\alpha}(-\lambda_k s^\alpha)| ds = \int_0^\eta s^{\alpha-1} E_{\alpha,\alpha}(-\lambda_k s^\alpha) ds$$

$$= -\frac{1}{\lambda_k} \int_0^\eta \frac{d}{ds} E_{\alpha,1}(-\lambda_k s^\alpha) ds = \frac{1}{\lambda_k} (1 - E_{\alpha,1}(-\lambda_k \eta^\alpha)). \qquad (4.50)$$

Here we recall: since $E_{\alpha,1}(z)$ is an entire function in z and $\Gamma(\alpha n + 1) = \alpha n \Gamma(\alpha n)$, we have

$$\frac{d}{ds} E_{\alpha,1}(-\lambda_k s^\alpha) = \frac{d}{ds} \left(\sum_{n=0}^\infty \frac{(-\lambda_k)^n s^{n\alpha}}{\Gamma(\alpha n + 1)} \right) = \sum_{n=1}^\infty \frac{(-\lambda_k)^n n \alpha s^{n\alpha-1}}{\Gamma(\alpha n + 1)}$$

$$= \sum_{n=1}^\infty \frac{(-\lambda_k)^n s^{n\alpha-1}}{\Gamma(\alpha n)} = s^{\alpha-1} (-\lambda_k) \sum_{m=0}^\infty \frac{(-\lambda_k)^m s^{m\alpha}}{\Gamma(\alpha m + \alpha)} = -\lambda_k s^{\alpha-1} E_{\alpha,\alpha}(-\lambda_k s^\alpha).$$

By $F \in L^2(0, T; H^{-1}(\Omega))$, we have

$$\langle F(\cdot, s), \varphi_k \rangle :=_{H^{-1}(\Omega)} \langle F(\cdot, s), \varphi_k \rangle_{H_0^1(\Omega)}$$

$$= (A^{-\frac{1}{2}} F(\cdot, s), A^{\frac{1}{2}} \varphi_k) = \lambda_k^{\frac{1}{2}} (A^{-\frac{1}{2}} F(\cdot, s), \varphi_k) \in L^2(0, T).$$

Therefore Proposition 2.3 yields

$$w_k(t) := \int_0^t \langle F(\cdot, s), \varphi_k \rangle (t - s)^{\alpha-1} E_{\alpha,\alpha}(-\lambda_k(t - s)^\alpha) ds \in H_\alpha(0, T),$$

and so (2.26) implies

$$\partial_t^\alpha w_k(t) = \frac{d}{dt}(J^{1-\alpha} w_k)(t), \quad t > 0. \tag{4.51}$$

Since $E_{\alpha,\alpha}(\eta)$ is an entire function in η, using the Lebesgue theorem, we can exchange $\int_0^t \left(\int_0^s \cdots d\xi \right) ds$ and $\sum_{m=0}^\infty$, so that

$$J^{1-\alpha} w_k(t) = \frac{1}{\Gamma(1 - \alpha)} \int_0^t (t - s)^{-\alpha} w_k(s) ds$$

$$= \frac{1}{\Gamma(1 - \alpha)} \int_0^t (t - s)^{-\alpha} \left(\int_0^s \langle F(\cdot, \xi), \varphi_k \rangle (s - \xi)^{\alpha-1} E_{\alpha,\alpha}(-\lambda_k(s - \xi)^\alpha) d\xi \right) ds$$

$$= \frac{1}{\Gamma(1 - \alpha)} \int_0^t (t - s)^{-\alpha} \left(\int_0^s \langle F(\cdot, \xi), \varphi_k \rangle (s - \xi)^{\alpha-1} \sum_{m=0}^\infty \frac{(-\lambda_k)^m (s - \xi)^{\alpha m}}{\Gamma(\alpha m + \alpha)} d\xi \right) ds$$

$$= \sum_{m=0}^\infty \frac{1}{\Gamma(1 - \alpha)} \int_0^t (t - s)^{-\alpha} \left(\int_0^s \langle F(\cdot, \xi), \varphi_k \rangle \frac{(-\lambda_k)^m (s - \xi)^{\alpha m + \alpha - 1}}{\Gamma(\alpha m + \alpha)} d\xi \right) ds.$$

Moreover, we exchange the orders of the integrals:

$$\frac{1}{\Gamma(1 - \alpha)} \int_0^t (t - s)^{-\alpha} \left(\int_0^s \langle F(\cdot, \xi), \varphi_k \rangle \frac{(-\lambda_k)^m (s - \xi)^{\alpha m + \alpha - 1}}{\Gamma(\alpha m + \alpha)} d\xi \right) ds$$

$$= \frac{1}{\Gamma(1 - \alpha)} \int_0^t \langle F(\cdot, \xi), \varphi_k \rangle (-\lambda_k)^m \left(\int_\xi^t \frac{(t - s)^{-\alpha} (s - \xi)^{\alpha m + \alpha - 1}}{\Gamma(\alpha m + \alpha)} ds \right) d\xi$$

$$= \frac{1}{\Gamma(1 - \alpha)} \int_0^t \langle F(\cdot, \xi), \varphi_k \rangle (-\lambda_k)^m \frac{\Gamma(1 - \alpha)\Gamma(\alpha m + \alpha)}{\Gamma(\alpha m + \alpha)} \frac{1}{\Gamma(\alpha m + 1)} (t - \xi)^{\alpha m} d\xi.$$

Therefore

$$J^{1-\alpha} w_k(t) = \sum_{m=0}^\infty \int_0^t \frac{(-\lambda_k(t - \xi)^\alpha)^m}{\Gamma(\alpha m + 1)} \langle F(\cdot, \xi), \varphi_k \rangle d\xi$$

$$= \int_0^t E_{\alpha,1}(-\lambda_k(t - \xi)^\alpha) \langle F(\cdot, \xi), \varphi_k \rangle d\xi.$$

By (4.51), we have

$$\partial_t^\alpha w_k(t) = \frac{d}{dt} J^{1-\alpha} w_k(t)$$

$$= \langle F(\cdot, t), \varphi_k \rangle + \int_0^t \frac{d}{dt}(E_{\alpha,1}(-\lambda_k(t-\xi)^\alpha)\langle F(\cdot, \xi), \varphi_k \rangle d\xi$$

$$= \langle F(\cdot, t), \varphi_k \rangle - \lambda_k \int_0^t (t-s)^{\alpha-1} E_{\alpha,\alpha}(-\lambda_k(t-s)^\alpha)\langle F(\cdot, s), \varphi_k \rangle ds = \langle F(\cdot, t), \varphi_k \rangle - \lambda_k w_k(t).$$

Hence

$$\partial_t^\alpha v_2(x, t) = \sum_{k=1}^\infty \langle F(\cdot, t), \varphi_k \rangle \varphi_k(x)$$

$$+ \left[-\sum_{k=1}^\infty \lambda_k \left(\int_0^t (t-s)^{\alpha-1} E_{\alpha,\alpha}(-\lambda_k(t-s)^\alpha)\langle F(\cdot, s), \varphi_k \rangle ds \right) \varphi_k \right] =: S_1 + S_2.$$

$$(4.52)$$

Therefore

$$\|\partial_t^\alpha v_2(\cdot, t)\|_{H^{-1}(\Omega)}^2 = \sum_{k=1}^\infty \frac{1}{\lambda_k} |\langle F(\cdot, t), \varphi_k \rangle|^2 + \|S_2(\cdot, t)\|_{H^{-1}(\Omega)}^2$$

$$\leq \|F(\cdot, t)\|_{H^{-1}(\Omega)}^2 + \|S_2(\cdot, t)\|_{H^{-1}(\Omega)}^2.$$

Next

$$\|S_2(\cdot, t)\|_{H^{-1}(\Omega)}^2 = \sum_{k=1}^\infty \frac{1}{\lambda_k}(S_2(\cdot, t), \varphi_k)^2$$

$$= \sum_{k=1}^\infty \lambda_k \left| \int_0^t (t-s)^{\alpha-1} E_{\alpha,\alpha}(-\lambda_k(t-s)^\alpha)\langle F(\cdot, s), \varphi_k \rangle ds \right|^2$$

and the Young inequality for the convolution (Lemma A.1) yields

$$\|S_2\|_{L^2(0,T;H^{-1}(\Omega))}^2$$

$$= \sum_{k=1}^\infty \lambda_k \int_0^T \left| \int_0^t (t-s)^{\alpha-1} E_{\alpha,\alpha}(-\lambda_k(t-s)^\alpha)\langle F(\cdot, s), \varphi_k \rangle ds \right|^2 dt$$

$$\leq \sum_{k=1}^\infty \lambda_k^2 \left(\int_0^T |s^{\alpha-1} E_{\alpha,\alpha}(-\lambda_k s^\alpha)| ds \right)^2 \frac{1}{\lambda_k} \int_0^T |\langle F(\cdot, s), \varphi_k \rangle|^2 ds.$$

Here (4.49) and (4.50) imply

$$\lambda_k^2 \left(\int_0^T |s^{\alpha-1} E_{\alpha,\alpha}(-\lambda_k s^\alpha)| ds \right)^2 = (1 - E_{\alpha,1}(-\lambda_k T^\alpha))^2 \leq 1. \tag{4.53}$$

Consequently,

$$\|S_2\|_{L^2(0,T;H^{-1}(\Omega))}^2 \leq \int_0^T \sum_{k=1}^\infty \frac{1}{\lambda_k} |\langle F(\cdot,s), \varphi_k \rangle|^2 ds$$

$$= \int_0^T \|F(\cdot,s)\|_{H^{-1}(\Omega)}^2 ds = \|F\|_{L^2(0,T;H^{-1}(\Omega))}^2. \tag{4.54}$$

Hence

$$\|\partial_t^\alpha v_2\|_{L^2(0,T;H^{-1}(\Omega))} \leq C \|F\|_{L^2(0,T;H^{-1}(\Omega))}.$$

With (4.46) and (4.48), since

$$v(x,t) - a(x) = v(x,t) - \sum_{k=1}^\infty (a, \varphi_k)\varphi_k = (v_1(x,t) - a(x)) + v_2(x,t),$$

we see that

$$\|\partial_t^\alpha (v - a)\|_{L^2(0,T;H^{-1}(\Omega))} \leq C(\|a\|_{L^2(\Omega)} + \|F\|_{L^2(0,T;H^{-1}(\Omega))}).$$

Fourth Step Here we verify that $v \in L^2(0,T;H_0^1(\Omega))$ and estimate $\|A^{\frac{1}{2}} v\|_{L^2(0,T;L^2(\Omega))}$, which is equivalent to $\|\nabla v\|_{L^2(0,T;L^2(\Omega))}$.

First Lemma 2.5 (i) yields

$$\left\| \sum_{k=1}^\infty A^{\frac{1}{2}}(a, \varphi_k) E_{\alpha,1}(-\lambda_k t^\alpha)\varphi_k \right\|_{L^2(\Omega)}^2 = \sum_{k=1}^\infty \lambda_k |(a, \varphi_k)|^2 |E_{\alpha,1}(-\lambda_k t^\alpha)|^2$$

$$\leq C \sum_{k=1}^\infty |(a, \varphi_k)|^2 \lambda_k \left(\frac{1}{1 + \lambda_k t^\alpha} \right)^2 = C t^{-\alpha} \sum_{k=1}^\infty |(a, \varphi_k)|^2 \left(\frac{\lambda_k^{\frac{1}{2}} t^{\frac{\alpha}{2}}}{1 + \lambda_k t^\alpha} \right)^2$$

$$\leq C t^{-\alpha} \|a\|_{L^2(\Omega)}^2.$$

Therefore, by $0 < \alpha < 1$, we obtain

$$\left\| \sum_{k=1}^{\infty} A^{\frac{1}{2}} (a, \varphi_k) E_{\alpha,1}(-\lambda_k t^{\alpha}) \varphi_k \right\|_{L^2(0,T;L^2(\Omega))}^2$$

$$\leq C \left(\int_0^T t^{-\alpha} dt \right) \|a\|_{L^2(\Omega)}^2 = C \|a\|_{L^2(\Omega)}^2.$$

In the same way as for the estimation of $\|S_2\|_{L^2(0,T;H^{-1}(\Omega))}^2$, using Lemma A.1, (4.53) and (4.54), we obtain

$$\left\| A^{\frac{1}{2}} \left(\sum_{k=1}^{\infty} \left(\int_0^t \langle F(\cdot, s), \varphi_k \rangle (t-s)^{\alpha-1} E_{\alpha,\alpha}(-\lambda_k(t-s)^{\alpha}) ds \right) \varphi_k \right) \right\|_{L^2(0,T;L^2(\Omega))}^2$$

$$= \sum_{k=1}^{\infty} \lambda_k \int_0^T \left| \int_0^t \langle F(\cdot, s), \varphi_k \rangle (t-s)^{\alpha-1} E_{\alpha,\alpha}(-\lambda_k(t-s)^{\alpha}) ds \right|^2 dt$$

$$\leq C \|F\|_{L^2(0,T;H^{-1}(\Omega))}.$$

Thus v satisfies (4.4), (4.5) and the same estimate as (4.7).

Finally we prove that v satisfies (4.3). For each $N \in \mathbb{N}$, we set

$$a_N(x) = \sum_{k=1}^{N} (a, \varphi_k) \varphi_k(x), \quad v_1^N(x, t) = \sum_{k=1}^{N} (a, \varphi_k) E_{\alpha,1}(-\lambda_k t^{\alpha}) \varphi_k(x),$$

$$v_2^N(x, t) = \sum_{k=1}^{N} \left(\int_0^t \langle F(\cdot, s), \varphi_k \rangle (t-s)^{\alpha-1} E_{\alpha,\alpha}(-\lambda_k(t-s)^{\alpha}) ds \right) \varphi_k(x)$$

and

$$v^N(x, t) = (v_1^N + v_2^N)(x, t).$$

Similarly to (4.47) and (4.52), we have

$$\partial_t^{\alpha} (v_1^N - a_N)(x, t) = -\sum_{k=1}^{N} \lambda_k (a, \varphi_k) E_{\alpha,1}(-\lambda_k t^{\alpha}) \varphi_k(x)$$

$$= -A \left(\sum_{k=1}^{N} (a, \varphi_k) E_{\alpha,1}(-\lambda_k t^{\alpha}) \varphi_k(x) \right) = -A v_1^N$$

and

$$\partial_t^\alpha v_2^N (x, t) = \sum_{k=1}^{N} \langle F(\cdot, t), \varphi_k \rangle \varphi_k$$

$$- A \left(\sum_{k=1}^{N} \int_0^t \langle F(\cdot, s), \varphi_k \rangle (t - s)^{\alpha-1} E_{\alpha,\alpha}(-\lambda_k (t - s)^\alpha) ds \varphi_k \right)$$

$$= - A v_2^N (x, t) + \sum_{k=1}^{N} \langle F(\cdot, t), \varphi_k \rangle \varphi_k.$$

Hence

$$\partial_t^\alpha (v_1^N - a)(x, t) + A v^N = \sum_{k=1}^{N} \langle F(\cdot, t), \varphi_k \rangle \varphi_k$$

for each $N \in \mathbb{N}$. Since

$$\left\| F - \sum_{k=1}^{N} \langle F(\cdot, t), \varphi_k \rangle \varphi_k \right\|_{L^2(0,T;H^{-1}(\Omega))}^2 = \int_0^T \sum_{k=N+1}^{\infty} |\langle F(\cdot, t), \varphi_k \rangle|^2 \lambda_k^{-1} dt \longrightarrow 0$$

as $N \longrightarrow \infty$ by $F \in L^2(0, T; H^{-1}(\Omega))$, we see that

$$\lim_{N \to \infty} (\partial_t^\alpha (v^N - a) + A v^N) = F \quad \text{in } L^2(0, T; H^{-1}(\Omega)).$$

We have already proved that v satisfies (4.7), so that

$$\lim_{N \to \infty} (\partial_t^\alpha (v^N - a) + A v^N) = \partial_t^\alpha (v - a) + A v \quad \text{in } L^2(0, T; H^{-1}(\Omega)),$$

which yields that v satisfies (4.3).   ∎

Chapter 5
Decay Rate as $t \to \infty$

5.1 Main Result

We consider

$$
\begin{cases}
\partial_t^\alpha (u - a) + A(t)u(x, t) = 0 & \text{in } L^2(\Omega),\ 0 < t < T, \\
u(\cdot, t) \in H_0^1(\Omega), & 0 < t < T, \\
u - a \in H_\alpha(0, T; L^2(\Omega)).
\end{cases}
\tag{5.1}
$$

Here

$$
-(A(t)v)(x, t) = \sum_{i,j=1}^{n} \partial_i (a_{ij}(x, t)\partial_j v(x)) + c(x, t)v,
\tag{5.2}
$$

where

$$
c \leq 0 \quad \text{on } \overline{\Omega} \times [0, \infty).
$$

In the case of $\alpha = 1$, we can prove that there exist constants $C > 0$ and $\theta > 0$ such that

$$
\|u(\cdot, t)\|_{L^2(\Omega)} \leq Ce^{-\theta t} \|a\|_{L^2(\Omega)}, \quad t > 0
$$

for each $a \in L^2(\Omega)$ by the classical energy estimate. For $0 < \alpha < 1$, we know that

$$
\|u(\cdot, t)\|_{L^2(\Omega)} \leq \frac{C}{t^\alpha} \|a\|_{L^2(\Omega)}, \quad t > 0
\tag{5.3}
$$

© The Author(s), under exclusive licence to Springer Nature Singapore Pte Ltd. 2020
A. Kubica et al., *Time-Fractional Differential Equations*, SpringerBriefs
in Mathematics, https://doi.org/10.1007/978-981-15-9066-5_5

in the case where the coefficients of $A(t)$ are independent of t (e.g., [26]). On the other hand, Vergara and Zacher [28] proved (5.3) for t-dependent $A(t)$. For t-dependent $A(t)$, in a conventional energy estimate, we multiply the equation by $u - a$ and integrate over Ω and we may obtain only a weak estimate

$$\|u(\cdot, t)\|_{L^2(\Omega)} \le \frac{C}{t^{\frac{\alpha}{2}}} \|a\|_{H_0^1(\Omega)}, \quad t > 0.$$

This is weaker than we can expect in terms of (5.3) for t-independent $A(t)$, because the decay rate is $\frac{\alpha}{2}$.

In this chapter, we modify the arguments in Vergara and Zacher [28] to prove (5.3) within our framework of solutions in the case where the coefficients of $A(t)$ are dependent on time t. More precisely, we prove:

Theorem 5.1 *In (5.2), we assume*

$$a_{ij} = a_{ji} \in C^2([0, \infty); C^1(\overline{\Omega})), \quad 1 \le i, j \le n, \tag{5.4}$$

$c \in L^\infty(\Omega \times (0, \infty))$, *and there exists a constant* $\mu_0 > 0$ *such that*

$$\sum_{i,j=1}^{n} a_{ij}(x, t)\xi_i\xi_j \ge \mu_0 \sum_{j=1}^{n} \xi_j^2, \quad x \in \overline{\Omega}, \, t \ge 0, \, \xi_1, ..., \xi_n \in \mathbb{R} \tag{5.5}$$

and

$$c(x, t) \le 0, \quad x \in \overline{\Omega}, \, t \ge 0. \tag{5.6}$$

Then there exists a constant $C > 0$ *such that*

$$\|u(\cdot, t)\|_{L^2(\Omega)} \le Ct^{-\alpha} \|a\|_{L^2(\Omega)}$$

for each solution u to (5.1).

For the proof, we have to return to the construction of the solution to (5.1) by the Galerkin approximation. By means of a conventional energy estimate which can be obtained by multiplying the fractional equation by the solution u to integrate over Ω by parts, we cannot reach the desired estimate $\|u(\cdot, t)\|_{L^2(\Omega)} \le Ct^{-\alpha} \|u(\cdot, 0)\|_{L^2(\Omega)}$. Thus we need another kind of energy inequality, namely Lemma 5.1 by Vergara and Zacher [28]. The lemma is technical and is shown in Sect. 5.2. Next, repeating the argument by the Galerkin approximation based on the lemma, we can complete the proof of the desired decay estimate.

5.2 Key Lemmata

The proof is based on another kind of energy inequality:

Lemma 5.1 *We set*

$$G(s) = \frac{1}{2}s^2 - s + \frac{1}{2}.$$

We assume that $u \in C(\overline{\Omega} \times [0, T])$,

$$u(x, \cdot) \in W^{1,1}(0, T) \quad for \, x \in \Omega, \tag{5.7}$$

and

$$t^{1-\alpha} \partial_t u \in L^\infty(\Omega \times (0, T)). \tag{5.8}$$

Then for each $t \in (0, T]$ *we have*

$$\begin{cases} d_t^\alpha \|u(\cdot, t)\|_{L^2(\Omega)}^2 + \frac{2\alpha}{\Gamma(1-\alpha)} \int_0^t (t-s)^{-\alpha-1} \|u(\cdot, t)\|_{L^2(\Omega)}^2 G\left(\frac{\|u(\cdot,s)\|_{L^2(\Omega)}}{\|u(\cdot,t)\|_{L^2(\Omega)}}\right) ds \\ + \frac{2t^{-\alpha}}{\Gamma(1-\alpha)} \|u(\cdot, t)\|_{L^2(\Omega)}^2 G\left(\frac{\|u(\cdot,0)\|_{L^2(\Omega)}}{\|u(\cdot,t)\|_{L^2(\Omega)}}\right) \le 2 \int_\Omega u(x, t) d_t^\alpha u(x, t) dx, \\ \qquad\qquad\qquad\qquad\qquad\qquad\qquad\qquad\qquad if \, \|u(\cdot, t)\|_{L^2(\Omega)} \ne 0, \\ d_t^\alpha \|u(\cdot, t)\|_{L^2(\Omega)}^2 \le 0, \quad if \, \|u(\cdot, t)\|_{L^2(\Omega)} = 0. \end{cases} \tag{5.9}$$

Furthermore

$$\|u(\cdot, t)\|_{L^2(\Omega)} d_t^\alpha \|u(\cdot, t)\|_{L^2(\Omega)} \le \int_\Omega u(x, t) d_t^\alpha u(x, t) dx. \tag{5.10}$$

Proof Choosing $M > 0$ sufficiently large, we have $u_M := u + M > 0$ on $\overline{\Omega} \times [0, T]$. We note that u_M satisfies (5.7) and (5.8). We have

$$2 \int_\Omega u_M(x, t) d_t^\alpha u_M(x, t) dx - d_t^\alpha \|u_M(\cdot, t)\|_{L^2(\Omega)}^2$$

$$= \frac{2}{\Gamma(1-\alpha)} \int_\Omega \int_0^t (t-s)^{-\alpha} (\partial_s u_M)(x, s)(u_M(x, t) - u_M(x, s)) ds dx$$

$$= \frac{2}{\Gamma(1-\alpha)} \int_\Omega |u_M(x, t)|^2 \left(\int_0^t (t-s)^{-\alpha} \partial_s \left(\frac{u_M(x, s)}{u_M(x, t)}\right) \left(1 - \frac{u_M(x, s)}{u_M(x, t)}\right) ds\right) dx.$$

Hence, since $G'(s) = s - 1$, we obtain

$$2 \int_{\Omega} u_M(x,t) d_t^{\alpha} u_M(x,t) dx - d_t^{\alpha} \|u_M(\cdot,t)\|_{L^2(\Omega)}^2$$

$$= -\frac{2}{\Gamma(1-\alpha)} \int_{\Omega} |u_M(x,t)|^2 \left(\int_0^t (t-s)^{-\alpha} \frac{\partial}{\partial s} \left(G\left(\frac{u_M(x,s)}{u_M(x,t)} \right) \right) ds \right) dx.$$

We integrate by parts to have

$$\int_0^t (t-s)^{-\alpha} \frac{\partial}{\partial s} \left(G\left(\frac{u_M(x,s)}{u_M(x,t)} \right) \right) ds = \lim_{h \to 0+} \int_0^{t-h} (t-s)^{-\alpha} \frac{\partial}{\partial s} \left(G\left(\frac{u_M(x,s)}{u_M(x,t)} \right) \right) ds$$

$$= \lim_{h \to 0+} -\alpha \int_0^{t-h} (t-s)^{-\alpha-1} G\left(\frac{u_M(x,s)}{u_M(x,t)} \right) ds + (t-s)^{-\alpha} G\left(\frac{u_M(x,s)}{u_M(x,t)} \right) \Big|_{s=0}^{s=t-h}$$

$$= \lim_{h \to 0+} -\alpha \int_0^{t-h} (t-s)^{-\alpha-1} G\left(\frac{u_M(x,s)}{u_M(x,t)} \right) ds + h^{-\alpha} G\left(\frac{u_M(x,t-h)}{u_M(x,t)} \right) - t^{-\alpha} G\left(\frac{u_M(x,0)}{u_M(x,t)} \right).$$

We shall show that

$$\lim_{h \to 0+} h^{-\alpha} |u_M(x,t)|^2 G\left(\frac{u_M(x,t-h)}{u_M(x,t)} \right) = 0, \quad t > 0. \tag{5.11}$$

Indeed, since

$$\left| h^{-\alpha} |u_M(x,t)|^2 G\left(\frac{u_M(x,t-h)}{u_M(x,t)} \right) \right| = \frac{h^{-\alpha}}{2} |u_M(x,t) - u_M(x,t-h)|^2,$$

assumption (5.8) yields

$$|u_M(x,t-h) - u_M(x,t)| = \left| \int_{t-h}^t \frac{\partial u_M}{\partial \xi}(x,\xi) d\xi \right|$$

$$\leq \int_{t-h}^t \left| \xi^{1-\alpha} \frac{\partial u_M}{\partial \xi}(x,\xi) \right| \xi^{\alpha-1} d\xi \leq C(t-h)^{\alpha-1} \int_{t-h}^t \|t^{1-\alpha} \partial_t u_M\|_{L^\infty(\Omega \times (0,T))} d\xi$$

$$\leq h(t-h)^{\alpha-1} \|t^{1-\alpha} \partial_t u_M\|_{L^\infty(\Omega \times (0,T))},$$

so that (5.11) holds.

Thus, letting $h \to 0+$, we obtain

$$2 \int_{\Omega} u_M(x,t) d_t^{\alpha} u_M(x,t) dx - d_t^{\alpha} \|u_M(\cdot,t)\|_{L^2(\Omega)}^2$$

$$= \frac{2\alpha}{\Gamma(1-\alpha)} \int_{\Omega} \int_0^t (t-s)^{-\alpha-1} |u_M(x,t)|^2 G\left(\frac{u_M(x,s)}{u_M(x,t)} \right) ds dx$$

$$+ \frac{2t^{-\alpha}}{\Gamma(1-\alpha)} \int_{\Omega} |u_M(x,t)|^2 G\left(\frac{u_M(x,0)}{u_M(x,t)} \right) dx.$$

Hence, by setting $\tau = 0$ or $= s$, the Cauchy–Schwarz inequality yields

$$\int_\Omega |u_M(x,t)|^2 G\left(\frac{u_M(x,\tau)}{u_M(x,t)}\right) dx$$

$$= \int_\Omega \left(\frac{1}{2}|u_M(x,t)|^2 + \frac{1}{2}|u_M(x,\tau)|^2 - u_M(x,t)u_M(x,\tau)\right) dx$$

$$\geq \frac{1}{2}\|u_M(\cdot,t)\|^2_{L^2(\Omega)} + \frac{1}{2}\|u_M(\cdot,\tau)\|^2_{L^2(\Omega)} - \|u_M(\cdot,t)\|_{L^2(\Omega)}\|u_M(\cdot,\tau)\|_{L^2(\Omega)}$$

$$= \|u_M(\cdot,t)\|^2_{L^2(\Omega)} G\left(\frac{\|u_M(\cdot,\tau)\|_{L^2(\Omega)}}{\|u_M(\cdot,t)\|_{L^2(\Omega)}}\right).$$

Thus (5.9) holds for u_M. If $\|u(\cdot,t)\|_{L^2(\Omega)} \neq 0$, then letting $M \to 0$, we obtain (5.9) for u. Next, we assume $\|u(\cdot,t)\|_{L^2(\Omega)} = 0$. Since $G(\eta) \geq 0$ for $\eta \geq 0$, inequality (5.9) for u_M yields

$$d_t^\alpha \|u(\cdot,t) + M\|^2_{L^2(\Omega)} \leq 2\int_\Omega (u(x,t) + M)d_t^\alpha (u(x,t) + M)dx$$

$$= 2\int_\Omega (u(x,t) + M)d_t^\alpha u(x,t)dx.$$

The right-hand side tends to 0 as $M \to 0$ by $\|u(\cdot,t)\|_{L^2(\Omega)} = 0$. Moreover,

$$d_t^\alpha \|u(\cdot,t) + M\|^2_{L^2(\Omega)} = \frac{1}{\Gamma(1-\alpha)}\int_0^t (t-s)^{-\alpha}\left(\frac{d}{ds}\int_\Omega (u(x,s) + M)^2dx\right) ds$$

$$= \frac{1}{\Gamma(1-\alpha)}\int_0^t (t-s)^{-\alpha}\left(\partial_s(\|u(\cdot,s)\|^2_{L^2(\Omega)}) + 2M\int_\Omega \partial_s u(x,s)dx\right) ds$$

$$= d_t^\alpha \|u(\cdot,t)\|^2_{L^2(\Omega)} + \frac{2M}{\Gamma(1-\alpha)}\int_0^t\int_\Omega (t-s)^{-\alpha}\partial_s u(x,s)dxds \longrightarrow d_t^\alpha \|u(\cdot,t)\|^2_{L^2(\Omega)}$$

as $M \to 0$ by $|\partial_s u(x,s)| \leq Cs^{\alpha-1}$ by (5.8). Hence $d_t^\alpha \|u(\cdot,t)\|^2_{L^2(\Omega)} \leq 0$, and (5.9) holds for $\|u(\cdot,t)\|_{L^2(\Omega)} = 0$.

Finally, we have to prove (5.10). We can assume that $\|u(\cdot,t)\|_{L^2(\Omega)} \neq 0$, because (5.10) is trivial if $\|u(\cdot,t)\|_{L^2(\Omega)} = 0$. By (5.8), we can directly verify $\frac{d}{ds}\|u(\cdot,s)\|_{L^2(\Omega)} \in L^1(0,T)$. First,

$$d_t^\alpha \|u(\cdot,t)\|^2_{L^2(\Omega)} - 2\|u(\cdot,t)\|_{L^2(\Omega)}d_t^\alpha \|u(\cdot,t)\|_{L^2(\Omega)}$$

$$= \frac{2}{\Gamma(1-\alpha)}\int_0^t (t-s)^{-\alpha}(\|u(\cdot,s)\|_{L^2(\Omega)} - \|u(\cdot,t)\|_{L^2(\Omega)})\frac{d}{ds}\|u(\cdot,s)\|_{L^2(\Omega)}ds$$

$$= \frac{2}{\Gamma(1-\alpha)}\int_0^t (t-s)^{-\alpha}\|u(\cdot,t)\|^2_{L^2(\Omega)}\left(\frac{\|u(\cdot,s)\|_{L^2(\Omega)}}{\|u(\cdot,t)\|_{L^2(\Omega)}} - 1\right)\frac{d}{ds}\left(\frac{\|u(\cdot,s)\|_{L^2(\Omega)}}{\|u(\cdot,t)\|_{L^2(\Omega)}}\right) ds$$

$$= \frac{2}{\Gamma(1-\alpha)} \int_0^t (t-s)^{-\alpha} \|u(\cdot,t)\|_{L^2(\Omega)}^2 \frac{d}{ds} \left[G\left(\frac{\|u(\cdot,s)\|_{L^2(\Omega)}}{\|u(\cdot,t)\|_{L^2(\Omega)}} \right) \right] ds$$

$$= -\frac{2\alpha}{\Gamma(1-\alpha)} \int_0^t (t-s)^{-\alpha-1} \|u(\cdot,t)\|_{L^2(\Omega)}^2 G\left(\frac{\|u(\cdot,s)\|_{L^2(\Omega)}}{\|u(\cdot,t)\|_{L^2(\Omega)}} \right) ds$$

$$+ \frac{2}{\Gamma(1-\alpha)} (t-s)^{-\alpha} \|u(\cdot,t)\|_{L^2(\Omega)}^2 G\left(\frac{\|u(\cdot,s)\|_{L^2(\Omega)}}{\|u(\cdot,t)\|_{L^2(\Omega)}} \right) \Bigg|_{s=0}^{s=t}$$

$$\geq \frac{-2\alpha}{\Gamma(1-\alpha)} \int_0^t (t-s)^{-\alpha-1} \|u(\cdot,t)\|_{L^2(\Omega)}^2 G\left(\frac{\|u(\cdot,s)\|_{L^2(\Omega)}}{\|u(\cdot,t)\|_{L^2(\Omega)}} \right) ds$$

$$- \frac{2t^{-\alpha}}{\Gamma(1-\alpha)} \|u(\cdot,t)\|_{L^2(\Omega)}^2 G\left(\frac{\|u(\cdot,0)\|_{L^2(\Omega)}}{\|u(\cdot,t)\|_{L^2(\Omega)}} \right). \text{ Here we used also (5.8). Hence}$$

$$d_t^\alpha \|u(\cdot,t)\|_{L^2(\Omega)}^2 + \frac{2\alpha}{\Gamma(1-\alpha)} \int_0^t (t-s)^{-\alpha-1} \|u(\cdot,t)\|_{L^2(\Omega)}^2 G\left(\frac{\|u(\cdot,s)\|_{L^2(\Omega)}}{\|u(\cdot,t)\|_{L^2(\Omega)}} \right) ds$$

$$+ \frac{2t^{-\alpha}}{\Gamma(1-\alpha)} \|u(\cdot,t)\|_{L^2(\Omega)}^2 G\left(\frac{\|u(\cdot,0)\|_{L^2(\Omega)}}{\|u(\cdot,t)\|_{L^2(\Omega)}} \right)$$

$$\geq 2\|u(\cdot,t)\|_{L^2(\Omega)} d_t^\alpha \|u(\cdot,t)\|_{L^2(\Omega)}.$$

Combining with (5.9), we reach (5.10). Thus the proof of Lemma 5.1 is complete. ∎

Moreover, we show a generalized extremum principle under weaker assumptions than Lemma 1.2.

Lemma 5.2 *Let* $f \in W^{1,1}(0,T)$ *attain its maximum over the interval* $[0,T]$ *at a point* $t_0 \in (0,T]$. *If for each* $\kappa \in (0,T)$, *there exists* $\beta \in (0,1]$ *such that* $f \in W^{1,\frac{1}{1-\beta}}(\kappa,T)$, *then* $(d_t^\alpha f)(t_0) \geq 0$ *for every* $\alpha \in (0,\beta)$.

Proof Firstly, we introduce the function $g(t) := f(t_0) - f(t)$ for $t \in [0,T]$. We notice that $g(t) \geq 0$ and $(d_t^\alpha g)(t) = -(d_t^\alpha f)(t)$ for $t \in [0,T]$. Then $g \in W^{1,1}(0,T)$ and $g(t_0) = 0$. Hence, for $\kappa \in (0,t_0)$, the Hölder inequality yields

$$|g(t)| \leq \int_t^{t_0} \left| \frac{dg}{ds}(s) \right| ds \leq \left\| \frac{dg}{dt} \right\|_{L^{\frac{1}{1-\beta}}(\kappa,T)} |t-t_0|^\beta \quad \text{for } t \in [\kappa, t_0]. \tag{5.12}$$

Therefore for small $h > 0$ satisfying $\kappa < t_0 - h < t_0$, we have

$$|g(t_0 - h)| \leq \left\| \frac{dg}{ds} \right\|_{L^{\frac{1}{1-\beta}}(\kappa,T)} h^\beta. \tag{5.13}$$

We note that $0 < \alpha < \beta$. For fixed $\alpha \in (0, \beta)$, we obtain

$$(d_t^\alpha g)(t_0) = \frac{1}{\Gamma(1-\alpha)} \int_0^\kappa (t_0 - s)^{-\alpha} \frac{dg}{ds}(s) ds + \frac{1}{\Gamma(1-\alpha)} \int_\kappa^{t_0} (t_0 - s)^{-\alpha} \frac{dg}{ds}(s) ds.$$

We fix $\varepsilon > 0$ arbitrarily. It follows from $g \in W^{1,1}(0, T)$ that $\frac{dg}{ds} \in L^1(0, T)$. By the Young inequality, we deduce that there exists sufficiently small $\kappa > 0$ such that the first integral is smaller than ε. As for the second one, integration by parts, $g \geq 0$ on $[0, T]$ and $g(t_0) = 0$ yield

$$\int_\kappa^{t_0} (t_0 - s)^{-\alpha} \frac{dg}{ds}(s) ds = \lim_{h \to 0+} \int_\kappa^{t_0-h} (t_0 - s)^{-\alpha} \frac{dg}{ds}(s) ds$$

$$= -(t_0 - \kappa)^{-\alpha} g(\kappa) + \lim_{h \to 0+} h^{-\alpha} g(t_0 - h) - \lim_{h \to 0+} \alpha \int_\kappa^{t_0-h} (t_0 - s)^{-\alpha-1} g(s) ds$$

$$\leq \lim_{h \to 0+} h^{-\alpha} g(t_0 - h) = 0.$$

For the last limit, we used (5.13) by noting $0 < \alpha < \beta$. Thus the proof of Lemma 5.2 is completed. ∎

5.3 Completion of Proof of Theorem 5.1

We apply the argument in Sect. 4.3 of Chap. 4 for proving the existence of u to (5.1), and we use the same notations. Since the non-homogeneous term $F(x, t)$ is identically zero, we need not introduce the approximation parameter $\varepsilon > 0$ like Sect. 4.3 of Chap. 4.

We recall that the function

$$u_N(x, t) := \sum_{\ell=1}^N p_{N,\ell}(t) \rho_\ell(x)$$

satisfies

$$\int_\Omega (d_t^\alpha u_N)(x, t) \rho_\ell(x) dx + \sum_{i,j=1}^N \int_\Omega a_{ij}(x, t) \partial_j u_N(x, t) \partial_i \rho_\ell(x) dx$$

$$= \int_\Omega c(x, t) u_N(x, t) \rho_\ell(x) dx$$

for $\ell = 1, ..., N$. In terms of (4.20), we apply Theorem 3.6 (iv) to see

$$p_{N,\ell} - c_\ell \in W_\alpha(0, T).$$

We multiply this identity by $p_{N,\ell}(t)$ and sum it up from $\ell = 1$ to N:

$$\int_\Omega u_N(x, t)(d_t^\alpha u_N)(x, t)dx + \sum_{i,j=1}^{N} \int_\Omega a_{ij}(x, t)(\partial_j u_N)(x, t)(\partial_i u_N)(x, t)dx$$

$$= \int_\Omega c(x, t)|u_N(x, t)|^2 dx.$$

Using (5.5) and (5.6), we obtain

$$\int_\Omega u_N(x, t)(d_t^\alpha u_N)(x, t)dx + \mu_0\|\nabla u_N(\cdot, t)\|_{L^2(\Omega)}^2 \le 0.$$

We define

$$u_{N,\eta} = u_N + \eta \rho_{N+1},$$

where $\eta > 0$ is a constant. Using the fact that $d_t^\alpha(\eta \rho_{N+1}) = 0$ and the orthogonality of $\{\rho_N\}_{N=1}^\infty$, we obtain

$$\int_\Omega u_{N,\eta}(x, t)d_t^\alpha u_{N,\eta}(x, t)dx + \mu_0\|\nabla u_N(t, \cdot)\|_{L^2(\Omega)}^2 \le 0.$$

By Theorem 3.6 (iv) in Chap. 3, we see that the functions $u_{N,\eta}$ satisfy the assumptions of Lemma 5.1, so that (5.10) and the Poincaré inequality yield

$$\|u_{N,\eta}(\cdot, t)\|_{L^2(\Omega)} d_t^\alpha \|u_{N,\eta}(\cdot, t)\|_{L^2(\Omega)} + c_0\|u_N(\cdot, t)\|_{L^2(\Omega)}^2 \le 0. \tag{5.14}$$

Here the constant $c_0 > 0$ is independent of N and η.

Since $u_{N,\eta} = u_N + \eta \rho_{N+1}$ and

$$(u_N, \rho_{N+1}) = \sum_{\ell=1}^{N} p_{N,\ell}(t)(\rho_\ell, \rho_{N+1}) = 0,$$

we have

$$\|u_{N,\eta}(\cdot, t)\|_{L^2(\Omega)}^2 = \|u_N(\cdot, t)\|_{L^2(\Omega)}^2 + \eta^2,$$

that is, $c_0 \|u_{N,\eta}(\cdot, t)\|_{L^2(\Omega)}^2 = c_0 \|u_N(\cdot, t)\|_{L^2(\Omega)}^2 + c_0 \eta^2$. Therefore

$$\|u_{N,\eta}(\cdot, t)\|_{L^2(\Omega)} d_t^\alpha \|u_{N,\eta}(\cdot, t)\|_{L^2(\Omega)} + c_0 \|u_{N,\eta}(\cdot, t)\|_{L^2(\Omega)}^2$$

$$= \|u_{N,\eta}(\cdot, t)\|_{L^2(\Omega)} d_t^\alpha \|u_{N,\eta}(\cdot, t)\|_{L^2(\Omega)} + c_0 \|u_N(\cdot, t)\|_{L^2(\Omega)}^2 + c_0 \eta^2 \le c_0 \eta^2$$

by (5.14). Consequently,

$$\|u_{N,\eta}(\cdot, t)\|_{L^2(\Omega)} d_t^\alpha \|u_{N,\eta}(\cdot, t)\|_{L^2(\Omega)} + c_0 \|u_{N,\eta}(\cdot, t)\|_{L^2(\Omega)}^2 \le \eta^2 c_0.$$

We note that $\|u_{N,\eta}(\cdot, t)\|_{L^2(\Omega)}^2 \ge \eta^2 > 0$, and we can divide by $\|u_{N,\eta}(\cdot, t)\|_{L^2(\Omega)}$ to obtain

$$d_t^\alpha \|u_{N,\eta}(\cdot, t)\|_{L^2(\Omega)} + c_0 \|u_{N,\eta}(\cdot, t)\|_{L^2(\Omega)} \le \eta c_0. \tag{5.15}$$

With the scalar product $(\cdot, \cdot)$ in $L^2(0, T)$, we have

$$(d_t^\alpha \|u_{N,\eta}(\cdot, t)\|_{L^2(\Omega)}, \xi) + (c_0 \|u_{N,\eta}(\cdot, t)\|_{L^2(\Omega)}, \xi) \le (\eta c_0, \xi) \tag{5.16}$$

for all $\xi \in C_0^\infty(0, T)$ satisfying $\xi \ge 0$. In view of (3.12), by Lemma 2.7 and integration by parts, we know that (2.51) holds for $u \in W^{1,1}(0, T)$ and $\xi \in C_0^\infty(0, T)$:

$$(d_t^\alpha \|u_{N,\eta}(\cdot, t)\|_{L^2(\Omega)}, \xi)$$

$$= (\|u_{N,\eta}(\cdot, t)\|_{L^2(\Omega)}, (d_t^\alpha)^* \xi) - \frac{1}{\Gamma(1-\alpha)} \left(\int_0^T t^{-\alpha} \xi(t) dt \right) \|u_{N,\eta}(\cdot, 0)\|_{L^2(\Omega)}.$$

By the definition of $u_{N,\eta}$, we can readily verify that $\|u_{N,\eta}(\cdot, t)\|_{L^2(\Omega)} \longrightarrow \|u_N(\cdot, t)\|_{L^2(\Omega)}$ in $L^1(0, T)$ and $\|u_{N,\eta}(\cdot, 0)\|_{L^2(\Omega)} \longrightarrow \|u_N(\cdot, 0)\|_{L^2(\Omega)}$ as $\eta \to 0$. Therefore

$$\lim_{\eta \to 0} (d_t^\alpha \|u_{N,\eta}(\cdot, t)\|_{L^2(\Omega)}, \xi) \tag{5.17}$$

$$= (\|u_N(\cdot, t)\|_{L^2(\Omega)}, (d_t^\alpha)^* \xi) - \frac{1}{\Gamma(1-\alpha)} \left(\int_0^T t^{-\alpha} \xi(t) dt \right) \|u_N(\cdot, 0)\|_{L^2(\Omega)}.$$

Since $\|u_N(\cdot, t)\|_{L^2(\Omega)} \in W^{1,1}(0, T)$ by Theorem 3.6 (iv), we apply (2.51) to the right-hand side of (5.17) for $\|u_N(\cdot, t)\|_{L^2(\Omega)}$, and we obtain

$$\lim_{\eta \to 0} (d_t^\alpha \|u_{N,\eta}(\cdot, t)\|_{L^2(\Omega)}, \xi) = (d_t^\alpha \|u_N(\cdot, t)\|_{L^2(\Omega)}, \xi).$$

Therefore, letting $\eta \to 0$ in (5.16), we reach

$$(d_t^\alpha \|u_N(\cdot, t)\|_{L^2(\Omega)} + c_0 \|u_N(\cdot, t)\|_{L^2(\Omega)}, \xi) \leq 0 \quad \xi \in C_0^\infty(0, T), \ \xi \geq 0.$$
$$(5.18)$$

We choose the mollifier χ_ε with $\varepsilon > 0$ defined by (3.19). We note

$$\theta(t) := d_t^\alpha \|u_N(\cdot, t)\|_{L^2(\Omega)} + c_0 \|u_N(\cdot, t)\|_{L^2(\Omega)} \in L^1(0, T).$$

Again using the letter θ, we denote the zero extension of θ to $\mathbb{R}$, and $\theta \in L^1(\mathbb{R})$. Setting $\xi(t) = \chi_\varepsilon(\tau - t)$ with arbitrarily fixed $\tau \in (0, T)$ and $\varepsilon > 0$, we see that $\xi \geq 0$ and $\xi \in C_0^\infty(0, T)$ with sufficiently small $\varepsilon > 0$. By (5.18) we obtain $\theta_\varepsilon(\tau) := \int_{-\infty}^\infty \theta(t)\chi_\varepsilon(\tau - t)dt \leq 0$ for small $\varepsilon > 0$. We know that $\theta_\varepsilon \longrightarrow \theta$ in $L^1(\mathbb{R})$ as $\varepsilon \to 0$ (e.g., [2]), so that we can choose a subsequence $\{\varepsilon_n\}_{n \in \mathbb{N}}$ with $\lim_{n\to\infty} \varepsilon_n = 0$ such that $\lim_{n\to\infty} \theta_{\varepsilon_n}(\tau) = \theta(\tau)$ for almost all $\tau \in (0, T)$. Hence, letting $n \to \infty$ in (5.18), we reach

$$d_t^\alpha \|u_N(\cdot, t)\|_{L^2(\Omega)} + c_0 \|u_N(\cdot, t)\|_{L^2(\Omega)} \leq 0 \quad \text{for almost all } t \in (0, T). \quad (5.19)$$

Using the power series of $E_{\alpha,1}(-c_0 t^\alpha)$, one can directly verify that

$$v(t) = \|u_N(\cdot, 0)\|_{L^2(\Omega)} E_{\alpha,1}(-c_0 t^\alpha)$$

satisfies

$$d_t^\alpha v(t) + c_0 v(t) = 0, \quad v(0) = \|u_N(\cdot, 0)\|_{L^2(\Omega)}.$$

Now we will apply Lemma 5.2 to obtain that

$$\|u_N(\cdot, t)\|_{L^2(\Omega)} \leq v(t) \quad \text{for every } t \in [0, T]. \quad (5.20)$$

Proof of (5.20) We set $f(t) := \|u_N(\cdot, t)\|_{L^2(\Omega)} - v(t)$. Then

$$d_t^\alpha f(t) + c_0 f(t) \leq 0, \quad t \in [0, T], \quad f(0) = 0. \quad (5.21)$$

It suffices to verify $f(t) \leq 0$ for $0 \leq t \leq T$. Assume contrarily that there exists $t_0 \in (0, T]$ such that $f(t_0) = \max_{t \in [0, T]} f(t) > 0$. From Theorem 3.6 (iv), we deduce that $f \in W^{1,\infty}(\kappa, T)$ for each $\kappa > 0$. Thus by Lemma 5.2, we have $(d_t^\alpha f)(t_0) \geq 0$, which means that $d_t^\alpha f(t_0) + c_0 f(t_0) > 0$. This is a contradiction with (5.21). Thus we have verified (5.20).

Next, let $\xi \in C_0^\infty(0, T), \geq 0$ be arbitrary. Then (5.20) yields

$$\int_0^T \xi(t) \|u_N(\cdot, t)\|_{L^2(\Omega)}^2 dt \leq \int_0^T \xi(t) v^2(t) dt.$$

There exists a subsequence $\{u_N\}$ (still indexed by N) such that $u_N \rightharpoonup u$ weakly in $L^2(0, T; H_0^1(\Omega))$, where u is the unique solution to (5.1) given by Theorem 4.1. Thus by the weak lower semi-continuity of the L^2-norm, we have

$$\int_0^T \xi(t)\|u(\cdot, t)\|_{L^2(\Omega)}^2 dt \leq \int_0^T \xi(t)v^2(t)dt$$

for each $\xi \in C_0^\infty(0, T)$, ≥ 0. Hence

$$\|u(\cdot, t)\|_{L^2(\Omega)} \leq v(t) \quad \text{for almost all } t \in [0, T].$$

Since $T > 0$ is arbitrary and the solution u is uniquely defined in $[0, \infty)$ by Theorem 4.1, we reach $\|u(\cdot, t)\|_{L^2(\Omega)} \leq v(t)$ for almost all $t \geq 0$, which means

$$\|u(\cdot, t)\|_{L^2(\Omega)} \leq \|a\|_{L^2(\Omega)} E_{\alpha,1}(-c_0 t^\alpha) \quad \text{for almost all } t \geq 0.$$

Finally, the asymptotic behavior of the Mittag-Leffler function implies

$$\|u(\cdot, t)\|_{L^2(\Omega)} \leq C_1 t^{-\alpha}\|a\|_{L^2(\Omega)}, \quad t > 0.$$

Thus the proof of Theorem 5.1 is complete.   ■

Chapter 6
Concluding Remarks on Future Works

The references on fractional differential equations are rapidly increasing and it is impossible to refer to them to some substantial extent. Thus in this book with limited pages, we concentrate on establishing one possible theory for the initial boundary value problems for time-fractional partial differential equations. Our theory redefines the time fractional derivative ∂_t^α keeping the consistency with the classical Riemann–Liouville derivative and the Caputo derivative, and ∂_t^α should be handled in a convenient way for the applications (e.g., the isomorphism between $H_\alpha(0, T)$ and $L^2(0, T)$).

We are obliged to give up comprehensive comparisons with other results. Also as for possible future studies, we are restricted to mention a few topics as follows.

1. In this book, we discuss only the case of the order $0 < \alpha < 1$ of the time-fractional derivative ∂_t^α. From the physical viewpoint, the case of $\alpha > 1$, in particular, $1 < \alpha < 2$ is important. The corresponding theory for the case $\alpha > 1$ can be constructed similarly; see Remark 2.4 of Chap. 2. However, we postpone the details to future works.

2 Here we do not discuss nonlinear equations

$$\partial_t^\alpha (u - a) + A(t)u = F(x, t, u, \nabla u).$$

Needless to say, one purpose of the linear theory developed in this book is the applications to the nonlinear theory.

3. Here we argue only the homogeneous Dirichlet boundary condition. For example, the Neumann boundary condition should be considered and initial nonhomogeneous boundary value problems are demanded. The theory for common partial differential equations such as $\alpha = 1, 2$ has been established satisfactorily (e.g., Lions and Magenes [19]). Such a theory is not only mathematically interesting but also is a basis for e.g., control problems, and widely applicable. As for non-homogeneous boundary value problems for fractional partial differential equations, we refer only to Yamamoto [29] and the references therein.

© The Author(s), under exclusive licence to Springer Nature Singapore Pte Ltd. 2020
A. Kubica et al., *Time-Fractional Differential Equations*, SpringerBriefs
in Mathematics, https://doi.org/10.1007/978-981-15-9066-5_6

4. There are various topics on inverse problems where we are required to determine parameters such as coefficients and orders of the derivatives of fractional differential equations by data of solutions which are over-determining in view of the well-posedness of initial boundary value problems. As for surveys, see the three chapters [16a–c] in the handbook edited by Kochubei, Luchko and Machado. The current book has established the foundations for future studies of inverse problems for fractional differential equations, and we can greatly expect the comprehensive development of studies of inverse problems.

Appendix A
Proofs of Two Inequalities

For convenience, we prove two inequalities which have been used in this book.

We set

$$(\rho * v)(t) = \int_0^t \rho(t-s)v(s)ds, \quad 0 < t < T.$$

Lemma A.1 (The Young Inequality for the Convolution) *Let* $1 \le p \le \infty$, $\rho \in L^1(0, T)$ *and* $v \in L^p(0, T)$. *Then*

$$\|\rho * v\|_{L^p(0,T)} \le \|\rho\|_{L^1(0,T)} \|v\|_{L^p(0,T)}.$$

Proof The proof for $p = \infty$ is straightforward. Let $p = 1$. Since

$$\int_0^T \left(\int_0^t \cdots ds \right) dt = \int_0^T \left(\int_s^T \cdots dt \right) ds \tag{A.1}$$

we have

$$\|\rho * v\|_{L^1(0,T)} = \int_0^T \left| \int_0^t \rho(t-s)v(s)ds \right| dt \le \int_0^T \left(\int_0^t |\rho(t-s)||v(s)|ds \right) dt$$

$$= \int_0^T \left(\int_s^T |\rho(t-s)|dt \right) |v(s)|ds = \int_0^T \left(\int_0^{T-s} |\rho(\eta)|d\eta \right) |v(s)|ds$$

$$\le \int_0^T \left(\int_0^T |\rho(\eta)|d\eta \right) |v(s)|ds = \|\rho\|_{L^1(0,T)} \|v\|_{L^1(0,T)}.$$

For the second equality from the last, we changed the integral variables $\eta = t - s$. Thus the proof in the case of $p = 1$ is complete.

© The Author(s), under exclusive licence to Springer Nature Singapore Pte Ltd. 2020
A. Kubica et al., *Time-Fractional Differential Equations*, SpringerBriefs
in Mathematics, https://doi.org/10.1007/978-981-15-9066-5

Finally, let $1 < p < \infty$. Let $q \in (1, \infty)$ satisfy $\frac{1}{p} + \frac{1}{q} = 1$, which is possible by $1 < p < \infty$. We have to estimate

$$\int_0^T \left| \int_0^t \rho(t-s)v(s) \right|^p dt.$$

Then

$$|\rho(t-s)v(s)| = |\rho(t-s)|^{\frac{1}{q}} \left(|\rho(t-s)|^{1-\frac{1}{q}} |v(s)| \right),$$

and the Hölder inequality yields

$$\int_0^t |\rho(t-s)||v(s)|ds = \int_0^t |\rho(t-s)|^{\frac{1}{q}} \left(|\rho(t-s)|^{1-\frac{1}{q}} |v(s)| \right) ds$$

$$\leq \left(\int_0^t |\rho(t-s)|ds \right)^{\frac{1}{q}} \left(\int_0^t |\rho(t-s)|^{p\frac{q-1}{q}} |v(s)|^p ds \right)^{\frac{1}{p}}$$

$$\leq \|\rho\|_{L^1(0,T)}^{\frac{1}{q}} \left(\int_0^t |\rho(t-s)||v(s)|^p ds \right)^{\frac{1}{p}}.$$

Therefore, using (A.1), we obtain

$$\int_0^T \left| \int_0^t \rho(t-s)v(s)ds \right|^p dt \leq \|\rho\|_{L^1(0,T)}^{\frac{p}{q}} \int_0^T \left(\int_0^t |\rho(t-s)||v(s)|^p ds \right) dt$$

$$= \|\rho\|_{L^1(0,T)}^{\frac{p}{q}} \int_0^T \left(\int_s^T |\rho(t-s)|dt \right) |v(s)|^p ds$$

$$= \|\rho\|_{L^1(0,T)}^{\frac{p}{q}} \int_0^T \left(\int_0^{T-s} |\rho(\eta)|d\eta \right) |v(s)|^p ds,$$

where we changed the integral variables $\eta = t - s$. Hence

$$\|\rho * v\|_{L^p(0,T)}^p \leq \|\rho\|_{L^1(0,T)}^{\frac{p}{q}} \|\rho\|_{L^1(0,T)} \|v\|_{L^p(0,T)}^p = \|\rho\|_{L^1(0,T)}^{\frac{p+q}{q}} \|v\|_{L^p(0,T)}^p,$$

which implies the conclusion by $\frac{1}{p} + \frac{1}{q} = 1$. Thus the proof of Lemma A.1 is complete.   ∎

Lemma A.2 (A Generalized Gronwall Inequality) *Let $C_0 > 0$ be a constant and $0 < \alpha < 1$. Moreover, let $r \in L^1(0, T), \geq 0$ in $(0, T)$. We assume that $u \in L^1(0, T)$ satisfies*

$$0 \leq u(t) \leq r(t) + C_0 \int_0^t (t-s)^{\alpha-1} u(s)ds, \quad 0 \leq t \leq T.$$

Then

$$u(t) \leq r(t) + C_1 e^{C_2 t} \int_0^t (t-s)^{\alpha-1} r(s) ds, \quad 0 \leq t \leq T. \tag{A.2}$$

Here the constants $C_1 > 0$ and $C_2 > 0$ are dependent on α, C_0, but independent of $T > 0$. We note that if $C_0 > 0$ is independent of $T > 0$, then (A.2) holds for $t > 0$ with $C_1, C_2 > 0$ which are independent of $t > 0$.

Proof We set

$$(Lr)(s) = C_0 \int_0^s (s-\xi)^{\alpha-1} r(\xi) d\xi, \quad 0 < s < T$$

for $r \in L^1(0, T)$. The assumption is written as

$$0 \leq u(t) \leq r(t) + (Lu)(t), \quad 0 < t < T. \tag{A.3}$$

Since $Lv \geq 0$ for $v \geq 0$ in $(0, T)$, we have

$$Lu \leq L(r + Lu) = Lr + L^2 u \quad \text{in } (0, T).$$

Then (A.3) yields

$$u \leq r + Lu \leq r + Lr + L^2 u \quad \text{in } (0, T).$$

We successively estimate to obtain

$$u \leq r + \sum_{k=1}^N L^k r + L^{N+1} u \quad \text{in } (0, T) \tag{A.4}$$

for each $N \in \mathbb{N}$. Exchanging the orders of the integrals, we have

$$(L^2 r)(t) = C_0^2 \int_0^t (t-s)^{\alpha-1} \left(\int_0^s (s-\xi)^{\alpha-1} r(\xi) d\xi \right) ds$$

$$= C_0^2 \int_0^t r(\xi) \left(\int_\xi^t (t-s)^{\alpha-1} (s-\xi)^{\alpha-1} ds \right) d\xi = \frac{C_0^2 \Gamma(\alpha)^2}{\Gamma(2\alpha)} \int_0^t (t-\xi)^{2\alpha-1} r(\xi) d\xi.$$

Here, by $r \in L^1(0, T)$ and the Young inequality, we note that all the functions including $r(\xi)$ in the calculations here are in $L^1(0, T)$. Continuing the calculations, we can reach

$$(L^k r)(t) = \frac{(C_0 \Gamma(\alpha))^k}{\Gamma(k\alpha)} \int_0^t (t-s)^{k\alpha-1} r(s) ds, \quad k \in \mathbb{N}. \tag{A.5}$$

We set $\theta = (C_0 \Gamma(\alpha))^{\frac{1}{\alpha}}$. Substitution of (A.5) into (A.4) yields

$$u(t) \le r(t) + \int_0^t \left(\sum_{k=1}^N \frac{\theta^{\alpha k}(t-s)^{\alpha k-1}}{\Gamma(\alpha k)} \right) r(s)ds + \frac{\theta^{\alpha(N+1)}}{\Gamma((N+1)\alpha)} \int_0^t (t-s)^{(N+1)\alpha-1} u(s)ds$$

$$= r(t) + \int_0^t \left(\sum_{k=1}^N \frac{\theta^{\alpha k} s^{\alpha k-1}}{\Gamma(\alpha k)} \right) r(t-s)ds + \frac{\theta^{\alpha(N+1)}}{\Gamma((N+1)\alpha)} \int_0^t (t-s)^{(N+1)\alpha-1} u(s)ds.$$

$$\text{(A.6)}$$

On the other hand, since $E_{\alpha,1}(z)$ is an entire function in $z \in \mathbb{C}$, we can verify

$$\frac{d}{ds} E_{\alpha,1}((\theta s)^\alpha) = \frac{d}{ds} \left(\sum_{k=0}^\infty \frac{(\theta s)^{\alpha k}}{\Gamma(\alpha k+1)} \right) = \sum_{k=1}^\infty \frac{\theta^{\alpha k}(\alpha k)s^{\alpha k-1}}{\Gamma(\alpha k+1)} = \sum_{k=1}^\infty \frac{\theta^{\alpha k} s^{\alpha k-1}}{\Gamma(\alpha k)}$$

$$= s^{\alpha-1}\theta^\alpha E_{\alpha,\alpha}((\theta s)^\alpha). \qquad \text{(A.7)}$$

Moreover, we have

$$|E_{\alpha,\alpha}(\eta)| \le C(1+\eta)^{\frac{1-\alpha}{\alpha}} e^{\eta^{\frac{1}{\alpha}}} + \frac{C}{1+\eta}$$

for all $\eta \ge 0$ (e.g., Theorem 1.5 (p. 35) in Podlubny [24]). Therefore

$$|E_{\alpha,\alpha}((\theta s)^\alpha)| \le C(1+\theta^\alpha s^\alpha)^{\frac{1-\alpha}{\alpha}} e^{\theta s} + \frac{C}{1+\theta^\alpha s^\alpha} \le C_1 e^{C_2 s}, \qquad s > 0,$$

where the constants $C, C_1 > 0$ are independent of s. Hence

$$\max_{0 \le s \le t} |E_{\alpha,\alpha}((\theta s)^\alpha)| \le C_1 e^{C_2 t}$$

for all $t \ge 0$. That is, (A.7) yields

$$\left| \sum_{k=1}^\infty \frac{\theta^{\alpha k}(t-s)^{\alpha k-1}}{\Gamma(\alpha k)} \right| \le C_1 e^{C_2 t}(t-s)^{\alpha-1}, \qquad 0 \le s \le t.$$

Letting $N \to \infty$ in (A.6), we obtain

$$u(t) \le r(t) + C_1 e^{C_2 t} \int_0^t (t-s)^{\alpha-1} r(s)ds$$

$$+ \lim_{N \to \infty} \frac{\theta^{\alpha(N+1)}}{\Gamma((N+1)\alpha)} \int_0^t (t-s)^{(N+1)\alpha-1} u(s)ds, \qquad 0 \le t \le T.$$

Let us choose $N_0 \in \mathbb{N}$ such that $(N_0 + 1)\alpha - 1 > 0$. Then for each $N \geq N_0$, we have

$$\left| \int_0^t (t - s)^{(N+1)\alpha - 1} u(s)ds \right| \leq t^{(N+1)\alpha - 1} \int_0^t |u(s)|ds \leq T^{(N+1)\alpha - 1} \|u\|_{L^1(0,T)}.$$

Consequently,

$$\left| \frac{\theta^{N+1}}{\Gamma((N+1)\alpha)} \int_0^t (t - s)^{(N+1)\alpha - 1} u(s)ds \right| \leq \theta T^{\alpha - 1} \frac{(\theta T^\alpha)^N}{\Gamma((N+1)\alpha)} \|u\|_{L^1(0,T)}$$

for each $t \in [0, T]$ and $N \geq N_0$. The Stirling formula yields

$$\lim_{N \to \infty} \frac{(\theta T^\alpha)^N}{\Gamma((N+1)\alpha)} = 0.$$

Thus the proof of Lemma A.2 is complete.   ■

Notation

$\| \cdot \|_{L^2(0,T)}, (\cdot, \cdot)$	L^2-norm, L^2-scalar product in $L^2(0, T)$
$\| \cdot \|_{L^2(\Omega)}, (\cdot, \cdot)_{L^2(\Omega)}$	L^2-norm, L^2-scalar product in $L^2(\Omega)$
$D_t^\alpha, 0 < \alpha < 1$	Riemann–Liouville derivative, see (1.3): $$D_t^\alpha u(t) = \frac{1}{\Gamma(1-\alpha)} \frac{d}{dt} \int_0^t (t-s)^{-\alpha} u(s) ds$$
$d_t^\alpha, 0 < \alpha < 1$	Caputo derivative, see (1.4): $$d_t^\alpha u(t) = \frac{1}{\Gamma(1-\alpha)} \int_0^t (t-s)^{-\alpha} \frac{du}{ds}(s) ds$$
$J^\alpha, \alpha > 0$	Riemann–Liouville fractional integral operator, see (1.5): $$J^\alpha f(t) = \frac{1}{\Gamma(\alpha)} \int_0^t (t-s)^{\alpha-1} f(s) ds$$
$\partial_t^\alpha, 0 < \alpha < 1$	Definition 2.1 in Sect. 2.3 of Chap. 2
$W^{1,1}(0, T), {}_0W^{1,1}(0, T)$	Definition (1.17)
$W_\alpha(0, T)$	Definition (2.5)
${}_0C^1[0, T], H_\alpha(0, T), H^\alpha(0, T),$ ${}_0H^\alpha(0, T), {}_0H^1(0, T)$	Section 2.2 of Chap. 2
$E_{\alpha,\beta}(z)$	Mittag-Leffler functions, see (2.28)
$\mathcal{D}(A)$	The domain of an operator A
$\mathcal{R}(A)$	The range of an operator A, i.e., $\mathcal{R}(A) = A\mathcal{D}(A)$
$\partial_i = \frac{\partial}{\partial x_i}, \partial_i^2 = \frac{\partial^2}{\partial x_i^2}, \partial_s = \frac{\partial}{\partial s}$	Partial derivatives
$\nabla = (\partial_1, \ldots, \partial_n)$	Gradient

© The Author(s), under exclusive licence to Springer Nature Singapore Pte Ltd. 2020
A. Kubica et al., *Time-Fractional Differential Equations*, SpringerBriefs
in Mathematics, https://doi.org/10.1007/978-981-15-9066-5

References

1. N.H. Abel, Résolution d'un problème de mécanique. J. Reine Angew. Math. **1**, 97–101 (1826)
2. R.A. Adams, *Sobolev Spaces* (Academic, New York, 1975)
3. A. Alsaedi, B. Ahmad, M. Kirane, A survey of useful inequalities in fractional calculus. Fract. Calc. Appl. Anal. **20**, 574–594 (2017)
4. E.G. Bajlekova, Fractional evolution equations in Banach spaces, Doctoral thesis, Eindhoven University of Technology, 2001
5. H. Brezis, *Functional Analysis, Sobolev Spaces and Partial Differential Equations* (Springer, Berlin, 2011)
6. K. Diethelm, *The Analysis of Fractional Differential Equations*. Lecture Note in Mathematics, vol. 2004 (Springer, Berlin, 2010)
7. L.C. Evans, *Partial Differential Equations* (American Mathematical Society, Providence, 1998)
8. R. Gorenflo, S. Vessella, *Abel Integral Equations*. Lecture Note in Mathematics, vol. 1461 (Springer, Berlin, 1991)
9. R. Gorenflo, M. Yamamoto, Operator theoretic treatment of linear Abel integral equations of first kind. Jpn. J. Ind. Appl. Math. **16**, 137–161 (1999)
10. R. Gorenflo, A.A. Kilbas, F. Mainardi, S.V. Rogosin, *Mittag-Leffler Functions, Related Topics and Applications* (Springer, Berlin, 2014)
11. R. Gorenflo, Y. Luchko, M. Yamamoto, Time-fractional diffusion equation in the fractional Sobolev spaces. Fract. Calc. Appl. Anal. **18**, 799–820 (2015)
12. B. Jin, B. Li, Z. Zhou, Subdiffusion with a time-dependent coefficient: analysis and numerical solution. Math. Comput. **88**, 2157–2186 (2019)
13. T. Kato, *Perturbation Theory for Linear Operators* (Springer, Berlin, 1980)
14. Y. Kian, M. Yamamoto, On existence and uniqueness of solutions for semilinear fractional wave equations. Fract. Calc. Appl. Anal. **20**, 117–138 (2017)
15. A.A. Kilbas, H.M. Srivastava, J.J. Trujillo, *Theory and Applications of Fractional Differential Equations* (Elsevier, Amsterdam, 2006)
16. A.N. Kochubei, Y. Luchko, J.A. Tenreiro Machado (eds.), *Handbook of Fractional Calculus with Applications*, vol. 2 (De Gruyter, Berlin, 2019); (a) Y. Liu, Z. Li, M. Yamamoto, Inverse problems of determining sources of the fractional partial differential equations, pp. 411–429; (b) Z. Li, Y. Liu, M. Yamamoto, Inverse problems of determining parameters of the fractional partial differential equations, pp. 431–442; (c) Z. Li, M. Yamamoto, Inverse problems of determining coefficients of the fractional partial differential equations, pp. 443–464
17. A. Kubica, M. Yamamoto, Initial-boundary value problems for fractional diffusion equations with time-dependent coefficients. Fract. Calc. Appl. Anal. **21**, 276–311 (2018)

18. Z. Li, X. Huang, M. Yamamoto, Initial-boundary value problems for multi-term time-fractional diffusion equations with x-dependent coefficients. Evol. Equ. Control Theory **9**, 153–179 (2020)

19. J.L. Lions, E. Magenes, *Non-homogeneous Boundary Value Problems and Applications*, vols. I, II (Springer, Berlin, 1972)

20. Y. Luchko, Maximum principle for the generalized time-fractional diffusion equation. J. Math. Anal. Appl. **351**, 218–223 (2009)

21. Y. Luchko, Some uniqueness and existence results for the initial-boundary value problems for the generalized time-fractional diffusion equation. Comput. Math. Appl. **59**, 1766–1772 (2010)

22. K.S. Miller, S.G. Samko, Completely monotonic functions. Integral Transforms Spec. Funct. **12**, 389–402 (2001)

23. A. Pazy, *Semigroups of Linear Operators and Applications to Partial Differential Equations* (Springer, Berlin, 1983)

24. I. Podlubny, *Fractional Differential Equations* (Academic, San Diego, 1999)

25. B. Ross, The development of fractional calculus 1695–1900. Hist. Math. **4**, 75–89 (1977)

26. K. Sakamoto, M. Yamamoto, Initial value/boundary value problems for fractional diffusion-wave equations and applications to some inverse problems. J. Math. Anal. Appl. **382**, 426–447 (2011)

27. H. Tanabe, *Equations of Evolution* (Pitman, London, 1979)

28. V. Vergara, R. Zacher, Optimal decay estimates for time-fractional and other nonlocal subdiffusion equation via energy methods. SIAM J. Math. Anal. **47**, 210–239 (2015)

29. M. Yamamoto, Weak solutions to non-homogeneous boundary value problems for time-fractional diffusion equations. J. Math. Anal. Appl. **460**, 365–381 (2018)

30. R. Zacher, Weak solutions of abstract evolutionary integro-differential equations in Hilbert spaces. Funkcial. Ekvac. **52**, 1–18 (2009)

Index

Printed in the United States
By Bookmasters